ALEX KIMMONS

INTRODUCTION
TO
CHEMISTRY

PREFACE

Chemistry is an interesting and fundamental branch of science. It gives us the chance to explain the secrets of nature.

What is water? What do we use in our cars as fuel? What is aspirin? What are perfumes made of? These kinds of questions and their answers are all part of the world of chemistry. Chemists work everyday to produce new compounds to make our lives easier with the help of this basic knowledge. All industries depend on chemical substances, including the petroleum, pharmaceutical, garment, aircraft, steel, electronics industries, etc.

This textbook is intended to help everyone understand nature. However, one does not need to be a chemist or scientist to comprehend the simplicity within the complexity around us.

The aim of our efforts was to write a modern, up-to-date textbook in which students and teachers could glean concise information about basic topics in chemistry.

"Introduction to Chemistry" is especially designed to introduce basic information about the unique facets of chemistry, matter, atoms, compounds, and mole concepts. An emphasis has been placed upon making the book student-friendly. Throughout the textbook, different figures, colorful tables, important reactions, funny cartoons, interesting extras, and reading passages have been added to help explain ideas.

We hope that after studying along with this book, you'll find chemistry in every part of your life.

CONTENTS

CHEMISTRY, A UNIQUE SCIENCE

INTRODUCTION TO CHEMISTRY

INTRODUCTION

Chemistry is LIFE!

Chemistry simply can be defined as ***the study of matter and its changes.*** This definition shows that the science of chemistry encompasses all substances in our life! In fact, ***chemistry is life!*** In this chapter, we try to explain why chemistry is so important in our lives. At the end, you'll see why chemistry is life.

Chemistry is a branch of natural sciences, such as physics, biology and geography.

Chemistry is UNIQUE!

As we hear the word ***science***, we should remember that it is ***the collective knowledge accumulated throughout the world's history.*** Although it's a branch of the natural sciences, ***chemistry is unique!***

Tree of natural sciences

Chemistry is in EVERYTHING!

Chemistry is unlike any other natural sciences because of one crucial difference; only chemistry is interrelated with all of the others. In other words, chemistry is related with all substances and objects. ***Chemistry is in everything!***

When we think about it, we can easily realize that ***there is nothing existing in our world*** that is not related to chemistry.

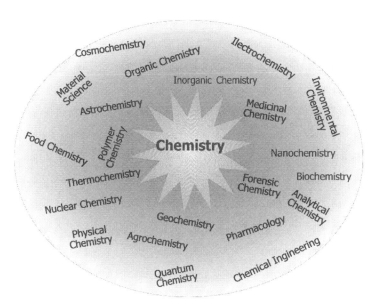

Chemistry is closely interweaved with all of the sciences.

1. WHY CHEMISTRY?

Because, chemists always synthesize or discover new substances and help technological developments. The modern residence you live in, the car your parents drive, the TV you watch, the Playstation you enjoy are all examples of the wonders achieved through the use of chemistry.

Because, chemists produce all modern or traditional herbal drugs and medicine; analyze blood, hormones, and urine; as well as diagnoze illnesses. This can only be possible through the use of chemistry.

Because, potable purified water, all organic or GM (Genetically Modified) fruits and vegetables are works of chemists.

Because, if we look around us, *chemistry is everywhere.* The water you drink, the air you breathe, the bread you eat, the CD you listen to or the clothes you wear, are all made out of chemicals!

Now, you see why chemistry is all around us.

Chemistry is EVERYWHERE!

All toys, including dolls and models, can be produced with the knowledge of chemistry.

We can decorate, paint, protect and clean our houses thanks to chemistry.

New polymers are making our life easier.

CHEMISTRY

This reading simply explains the importance of chemistry in our lives. You can see chemistry in every part of our life. For example, while preparing this textbook we used pencils, pens and papers. To type it, computers were used and printers to print its pages with different inks (paints); paper and printing machines were also utilized. All materials and machines used in the printing of this book are also the products of chemicals. In other words, they are produced with the help of different elements or/and compounds. All these are the subjects of chemistry. Yet another example, what if you were to become ill, how would your ailment be diagnozed?

Computer technology only thrives with the help of chemistry.

Chemistry even helps us with our clothing and footwear.

Long-term preservation of our packaged food is only made possible with chemicals.

Without chemicals, our body, for example, and our nervous system couldn't function.

Chemistry helps us in coloring our world.

IS LIFE

Doctors may guess the disfunction (illness) you are suffering from. In other words, the apparent chemical change that occured in your organs (in the tissues or cells). If a doctor diagnoses your illness, he will probably advice some medicine (products of chemicals) or different remedies (all contain different compounds or chemicals).

No chemistry! No transportation!

You cannot escape from the scope of chemistry because we are living in a material world, or chemical world. Have you ever thought that we need chemicals everyday in every second to live?

In fact, the air we breathe contains N_2, O_2, ... all are chemicals. The water we drink many times a day is also a chemical. Salt, pepper, sugar, ice-cream, all fruits and all vegetables (all foods) are all composed of chemicals!

All the drugs are produced from chemicals.

Paper and pencils are all produced with chemical materials.

All patients need the products of chemistry in diagnoses or treatments.

2. WHAT IS SCIENCE?

As you remember, chemistry is a branch of science. To understand this better, let's first see what science is. People observe their surroundings, other creatures and themselves to discover their nature and to find out their relationship among themselves. When they come face - to - face with problems, they use systematic methods to overcome them. From ancient times until today, acquired knowledge has been collected on a regular basis. This accumulated knowledge is known simply as **science.**

Throughout the world's history, humankind's knowledge has been continuously transmitted to newer generations. People shared this knowledge by writing, speaking and searching for the relations between the cause and effect of facts. As a result, we were able to reach conclusions. All of these activities, hypothethized carefully and tested systematically, are called **scientific studies.** Individuals who conduct scientific studies are called **scientists.**

Chemist : A scientist who studies chemistry.

Chemists are scientists.

Today's generation is grateful for past and present scientific studies since their welfare has been enhanced. Chemistry, as we've seen, is an important branch of science. It's certainly true that chemists who study the changes that occur in the structure of substances have a great contribution in the development of science and technology. The applications of scientific principles in the service of mankind is called **technology**.

Although developments in industry and technology enhance the welfare of human beings, there are negative sides of scientific developments. Some gases cause air pollution, poisonous chemical wastes and their by-products cause cancer; all weapons and atomic bombs threaten the balance of mankind.

At first glance, the branches of scientific studies seem to be boring due to groups of formulas, theories and definitions. But, actually, science reflects the attractive sides of the universe and informs us what is happening in our their surroundings.

Steps of Scientific Study

All scientific studies should be carried out in certain steps.

1. Observation

Observation, which is the first step of scientific study, is the collection of information gathered from the use of our senses. A subject can be observed through touch, taste, hearing, or smell. During observations, better results can be obtained by using instruments that increase the sensitivity of our senses. We must always keep in mind that the observation process is not right or wrong.

2. Hypothesis and (Research for Reasons of Regularities)

Hypothesis is the explanation of an occurence or a problem by using the knowledge obtained from previous observations.

All hypotheses may not always be true. For this reason, the following steps should be applied one by one to prove a hypothesis.

An observation in a lake

3. Experiment

The collection of actions and observations performed to verify or falsify a hypothesis or research. Experiments are generally performed in laboratories.

4. Sharing Results

The definition of science, as we mentioned before, is the accumulated knowledge that has been formed through centuries. If we share that knowledge with others, we cannot continuously start from the beginning of our scientific studies.

Today, telecommunication technologies have changed the world into a global village; hence, news can now be spread throughout the world in a matter of seconds through the use of the Internet. Similarly, scientific and technological development can easily be propagated through the means of books, scientific journals and the Internet.

In a science class

Computer technologies have changed the world into a global village.

The word **chemistry** officially comes from the old French **alkemia**, but has an even older root stemming from the Arabic origin of **al-kimia**, which means the art of transformation.

We can trace the beginning of chemistry to ancient times. The first chemists were mainly concerned with pottery, metallurgy, dyes, and food. We can retrace the earliest chemical principle to 3500 B.C. in Egypt and Mesopotamia.

Chemistry's beginning, or alchemy, as it was then known, was equal to performing magic or having superpowers. Alchemy was the practice of combining elements of chemistry, physics, religion, mysticism, astrology, art, signs, metallurgy, and medicine.

The most famous interests of alchemists were the transmutation of metals to gold, and the search for **Ab-u hayat**, or the elixir of life to produce immortality.

In the Middle Ages, Muslim scientists **Jabir bin Hayyan** - the first to use lab equipment - was known as Geber, or the **Father of Chemistry** in Europe; and Abu Bakr-Al-Razi (865-925); both greatly contributed in chemistry's early beginnings.

EA T

¾O
STATES

Greek thinkers believed that there were just four states of matter; water, fire, earth, and air (in some sources, the fifth was the cosmos).

Jabir bin Hayyan (721 - 815)
He is known as the Father of Chemistry.

Latin translations of these two Muslim scientists' discoveries helped build the fundamentals of chemistry. Six centuries later, European scientists **Robert Boyle** and **Antoine Lavoisier** - regarded as the **Fathers of Modern Chemistry** - built the basis for what we know today as modern chemistry

R. Boyle (1627-1691)   A. Lavoisier (1743-1794)

The last 200 years have heralded a multitude of scientific equipment and millions of chemical compounds which were all discovered through synthesis.

Of all the sciences, did you know that chemistry has the largest encyclopaedia of terminology?

Even laboratories have been altered dramatically through history.

In the 21st century, chemistry has become the largest collection of knowledge (science) because of its many sub - branches. Today, the world's largest and most current database that contains information regarding chemical substances, CAS, has more than 29 million chemical substances. Everyday, 4000 new substances are added to this database.

CAS = Chemical Abstracts Service

3. THE EXPERIMENTAL WORLD OF CHEMISTRY

The word *experiment* comes from Latin *experiri* or *experimentum*, which means *attempt*. In the scientific method, experiments are the processes from which all empirical knowledge is born.

3.1. CHEMISTRY IS AN EXPERIMENTAL SCIENCE

In chemistry, chemists always need experiments to research cause/effect relationships between phenomena.

Like other scientists, such as physicists, biologists, and many others, chemists must follow four systematic steps while experimenting:

> 1. *Setting up the experimental equipment and procedures*
> 2. *Conducting the experiment*
> 3. *Recording data*
> 4. *Analyzing results, reaching a conclusion and sharing results*

Lab. = Laboratory

Through the use of experiments, chemists understand the properties and changes which occur in matter and the various reactions between different substances.

3.2. WORKING IN A CHEMISTRY LABORATORY

Experiments should be performed in laboratories. A laboratory, often simply called a **lab**, is a specially designed environment to conduct experiments or scientific research.

Safety always comes first in lab!

While performing chemistry experiments in school labs, chemicals, water, fire, and glassware should be handled carefully.

Although laboratories are not playgrounds, they can be very enjoyable places. But, *safety always comes first;* sometimes small unexpected accidents may cause serious injuries. Therefore, students, teachers, and lab technicians should always use utmost caution in a lab environment.

Ensuring A Safe Laboratory

In a chemistry lab, these must be available:

a. Fire extinguisher   b. Fire blanket
c. Safety shower   d. Eyewash station
e. First aid kit   f. Container for sharp objects
g. Air conditioner   h. Telephone

To ensure that your school lab is a safe place, here are some basic safety rules about laboratory equipment and hazard warning symbols.

Safety Rules

Students must always obey the following safety rules in a school lab:

1. Always listen carefully to all instructions given by your chemistry teacher or lab technician before any experiment.

2. Learn all the safety rules necessary for your experiment.

3. Always wear a lab jacket (apron) and safety goggles (glasses).

4. Read all necessary information about the experiments.

5. Wear gloves when using heat, chemicals and glassware in experiments.

6. Do not throw any chemicals.

7. Do not touch chemicals with your fingers.

8. Tie long hair.

9. Do not horse around in a lab.

10. Quickly report all accidents to your teacher.

11. Do not eat, drink, smell or taste any chemical.

THE SAFE WAY
IS THE
BEST WAY

Safety always comes first in a lab

Laboratory Equipment

In chemistry experiments, different lab equipment can be used. Some of them appear below:

Do not use any lab equipment without permission!

Test Tube
(Mixing substances)

Erlenmeyer flask
(Toxic substances)

Beaker
(Used for measuring approximate volumes)

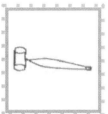

Test tube holder
(To hold hot test tubes)

Graduated cylinder
(To measure exact volumes of liquids)

Stands and clamps
(To hold and fix lab equipment)

Some equipment may cause fatal accidents!

Wash bottle
(Dilution)

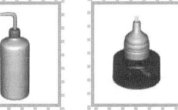

Burners
(For heating purposes)

Round bottom flask
(Mixing, collecting liquids)

Funnel
(Used for filtration or pouring liquids)

Test tube rack
(To park test tubes)

Tripod
(Used with burners)

Hazard Warning Symbols

Many chemicals are hazardous. Therefore, hazard warning symbols are placed on containers of harmful chemicals and on laboratory walls to explain and inform of possible dangers.

Hazard labels or warning symbols can be found in a lab, but they also can be seen in many other venues in life, such as on roadsides...

4. BASIC MEASUREMENTS AND CHEMISTRY

4.1. UNITS AND QUANTITIES

What is the time? How many kilos of potatoes? How many liters of milk? These are common questions asked in everyday's marketplace. All of these need answers using different quantities. Do you know how we can measure these quantities? To better answer these questions, let's first see what measurement is.

Measurement (or the act of measuring) means *a comparison of a desired quantity with a standard, or base quantity.*

In chemistry, many quantities - such as mass, volume, temperature, pressure, and energy - are commonly used.

In quantity measurements, we need to use certain standards. These standards are called **units.**

A unit is the amount of quantity used as a standard of measurement.

Some of these units in chemistry are as follows:

Mass: Amount of substance, shown as **m**. The common mass unit is a **gram** (g).

1 gram = 1000 milligram (mg)

1 gram = 0.001 kilogram (kg) $\Rightarrow$ (1000 gram = 1 kg)

Volume: Space occupied in the cosmos. The common volume unit is a **liter** (or litre) shown by **L**.

1 L = 1000 milliliter (mL) or 1 L = 1 dm^3 (cubic decimeter)

1 dm^3 = 1000 cm^3

Measurement is a comparison.

An electronic scale is used to measure mass.

Graduated cylinder

Volumetric flask

Temperature: Degree of heat or cold. The common temperature unit is a degree Celsius (°C). In some countries, the Fahrenheit Scale (F) is also used. The kelvin (K) is yet another unit of temperature. It is very important in chemistry because a kelvin is a base unit for temperature in the SI unit system. The conversion of °C to K is given below:

$$K = °C + 273$$

Celsius and Fahrenheit Scale Thermometer measures temperature.

Quantity	Symbol	Unit	Symbol
Length	L	meter	m
Mass	m	kilogram	kg
Time	t	second	s
Electric current	I	ampere	A
Temperature	T	kelvin	K
Amount of substance	N	mole	mol
Luminous intensity	I_v	candela	cd

Table 1: Basic SI units

SI is the abbreviation of "Systéme International d'Unités", or in English, the "International System of Units". The SI unit system was developed in 1960 (Table 1).

4.2. MULTIPLIERS AND UNIT CONVERSION

In chemistry, we use **Multiplying Prefixes** instead of using very small or very large numbers (Table 2).

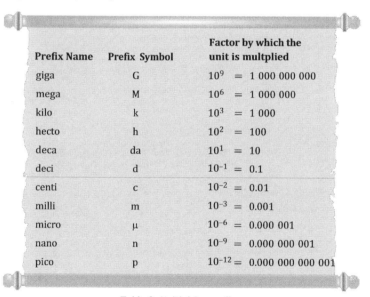

Prefix Name	Prefix Symbol	Factor by which the unit is multplied
giga	G	10^9 = 1 000 000 000
mega	M	10^6 = 1 000 000
kilo	k	10^3 = 1 000
hecto	h	10^2 = 100
deca	da	10^1 = 10
deci	d	10^{-1} = 0.1
centi	c	10^{-2} = 0.01
milli	m	10^{-3} = 0.001
micro	μ	10^{-6} = 0.000 001
nano	n	10^{-9} = 0.000 000 001
pico	p	10^{-12} = 0.000 000 000 001

Table 2: Multiplying prefixes

A stopwatch is used to measure time.

To change cubic centimeters to liters, use the ratio 1 L / 1000 cm³; to convert liters to cubic centimeters, use the ratio 1000 cm³ /1 L. In general, when you make a conversion, choose the factor that cancels out the initial unit.

> Desired quantity = Initial quantity · Conversion factor

Conversions between English and metric units can be made by using Table 3:

Metric	English	Metric - English
Mass		
$1 \text{ kg} = 10^3 \text{ g}$	1 lb = 16 oz	1 lb = 453.6 g
$1 \text{ mg} = 10^{-3} \text{ g}$	1 short ton = 2000 lb	1 g = 0.03527 oz
$1 \text{ ton} = 10^3 \text{ kg}$		1 ton = 1.102 short ton
Volume		
$1 \text{ m}^3 = 10^6 \text{ cm}^3 = 10^3 \text{ L}$	1 gallon = 4 qt	$1 \text{ ft}^3 = 28.32 \text{ L}$
$1 \text{ cm}^3 = 1 \text{ mL} = 10^{-3} \text{ L}$	1 qt (U.S.) = 57,75 in^3	1 L = 1.057 qt (U.S.)
	1 quart = 0.9464 L	
Length		
$1 \text{ km} = 10^3 \text{ m}$	1 ft = 12 in	1 in = 2.54 cm
$1 \text{ cm} = 10^{-2} \text{ m}$	1 yd = 3 ft	1 m = 39.37 in
$1 \text{ mm} = 10^{-3} \text{ m}$		1 mile = 1.609 km

Table 3: Conversions between English and metric units

qt : quart
lb : pound (libre)
ft : foot
yd : yard
oz : ounce

Example 1

What is the mass of 5 kg of sugar in the following units?

a. In grams

b. In milligrams

c. In tons

Solution

a. $\text{Sugar} = 5 \text{ kg} \cdot \dfrac{1000 \text{ g}}{1 \text{ kg}} = 5000 \text{ g}.$ $= 5,000,000 \text{ mg} = 5 \cdot 10^6 \text{ mg}$

b. $\text{Sugar} = 5 \text{ kg} \cdot \dfrac{1000 \text{ g}}{1 \text{ kg}} \cdot \dfrac{1000 \text{ mg}}{1 \text{ g}}$

c. $\text{Sugar} = 5 \text{ kg} \cdot \dfrac{0.001 \text{ ton}}{1 \text{ kg}} = 0.005 \text{ ton}$

Example 2

Calculate the volume of a swimming pool (2m · 5m · 10m).

a. in m^3

b. in dm^3

c. in mm^3

Solution

Volume of swimming pool becomes

a. $V = a \cdot b \cdot c = 2\,m \cdot 5\,m \cdot 10\,m \Rightarrow V = 100\,m^3$

b. $V = 100\ \cancel{m^3} \cdot \dfrac{1000\,dm^3}{1\,\cancel{m^3}} = 100000\,dm^3$

c. $V = 100\ \cancel{m^3} \cdot \dfrac{1000\ \cancel{dm^3}}{1\,\cancel{m^3}} \cdot \dfrac{1000\ \cancel{cm^3}}{1\,\cancel{dm^3}} \cdot \dfrac{1000\,mm^3}{1\,\cancel{cm^3}} = 10^{11}\,mm^3$

Example 3

According to a road sign, the distance from Istanbul to Ankara is about 295 miles. Express this distance in kilometers. (1 mile = 1.609 km)

Solution

1 mile = 1.609 km is conversion factor.

Desired quantity = Initial quantity · Conversion factor

Distance Istanbul - Ankara = 295 $\cancel{miles} \cdot \dfrac{1.609\,km}{1\,\cancel{mile}} = 475\,km$

A road sign informing distance.

Example 4

Convert the following temperature rules given in degree Celsius (°C) to the kelvin (K) scale.

a. Water freezes at 0 °C.

b. Melting point of copper is 1083 °C.

Solution

a. K = °C + 273 = 0 + 273 = 273 K

b. K = °C + 273 = 1083 + 273 = 1356 K

1. What is science?

2. What are the differences between chemistry and other branches of natural sciences?

3. Why is chemistry called a unique science?

4. Can you find anything which is not related to chemistry around you? Discuss the results with friends and your teacher.

5. What is an experiment?

6. Why do we obey safety rules in a school lab? Explain with examples.

7. Why do we need eye washers and safety showers in school laboratories?

8. Which types of labware should be found in a chemistry laboratory?

9. Compare glassware found in your kitchen to those found in a school chemistry laboratory.

10. How many calories can you get if you eat an orange (about 200 g) or drink a bottle of coke (0.25 L)? Research.

11. In daily life, what is the most common unit that you use?

12. Where and why do we need units of pressure? Research.

13. Find a medical thermometer and an outdoor thermometer. Compare the similarities and differences between them.

14. Calculate the area of your school basketball court in m^2, km^2 and mm^2.

15. Fill the blanks with the suitable signs (=, > or <)

   a. 0.1 meter...................................10 cm
   b. 2 square meters...................................200 cm^2
   c. 5 cubic millimeters ....$5 \cdot 10^{-3}$ cm^3
   d. 2 hour 10 minutes...................8340 sec

16. What should be the volume of the substance x in cm^3? (x is insoluble in water)

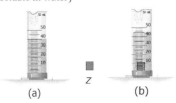

   (a)   Z   (b)

17. Match the following words.

   1. Temperature   a. Electronic scale
   2. Mass   b. Thermometer
   3. Volume   c. Stopwatch
   4. Length   d. Graduated cylinder
   5. Time   e. Ruler

18. Name the following glassware.

   a. ....   b. ....   c. ....

19. Where do we see the following words in daily life? Research

   a. Gigahertz (Ghz)   d. Kilogram (kg)
   b. Megabyte (MB)   e. Centiliter (cL)
   c. Megapixel (MP)   f. Microwave (MW).

20. Research the following:

   a. The volume of the Earth and the volume of the moon in m^3.

   b. The distance between the Sun and the Earth in meters.

21. Answer the following questions about a person's daily intake of water. (Hint: Accept that a year is 365 days and an average person needs 3 L water per day)

   a. How many tons of water would one need in 50 years?

   b. Convert the result found to liters. (Assume 1 kg water has 1 L volume).

   c. Does that amount of water (refer to answer 21.b) fill a small swimming pool with dimensions of 10 m · 4 m · 2 m?

MULTIPLE CHOICE QUESTIONS

1. Which of the following is not a natural science?

   A) Physics   B) Geology   C) History

        D) Chemistry   E) Astronomy

2. Why is chemistry called a unique science?

   A) Chemistry has many topics.
   B) Chemists know everything.
   C) Chemistry is related with everything around us.
   D) Chemists do different experiments everyday.
   E) Chemistry has branches such as organic, physical, analytical, etc...

3. Which of the following is not a step in a scientific study?

   A) Observation   B) Hypothesis   C) Experiment

        D) Reading   E) Evaluation

4. What were chemists once named?

   A) Old chemists   B) All chemists   C) Alchemists

        D) Only chemists   E) Early chemists

5. Give the right order for the steps in an experiment.

   I. Recording data
   II. Conducting experiment
   III. Setting up experimental equipment
   IV. Analyzing results

   A) II, I, III, IV   B) I, II, III, IV   C) III, II, IV, I

        D) II, IV, III, I   E) IV, III, II, I

6. Which of the following is expected from a chemistry student in a lab?

   A) Obeying safety rules
   B) Chewing gum
   C) Drinking liquids
   D) Playing with friends
   E) Eating sandwiches

7. Which is probably not found in a chemistry lab?

   A) Chemicals   B) Flasks   C) Beakers

        D) Compass   E) Burners

8. Which is not equal to 1000 mL?

   A) 1 L   B) 100 cL   C) 10 dL

        D) 0.01 hL   E) 0.01 daL

9. Which of the following unit is generally not used in a chemistry lab?

   A) Gram (g)
   B) Second (s)
   C) Degree Celsius (°C)
   D) Milliliter (mL)
   E) Light year

10. How many milligrams can be found in 5 tons?

    A) 500,000   B) 5,000,000   C) 50,000,000

         D) 500,000,000   E) 5,000,000,000

11. Which of the following classification of sciences does not exist?

    A) Tropical sciences
    B) Social sciences
    C) Physical sciences
    D) Life sciences
    E) Natural sciences

12. Which one of the following is not a unit of length?

    A) foot   B) kilogram   C) meter

         D) mile   E) inch

CRISSCROSS PUZZLE

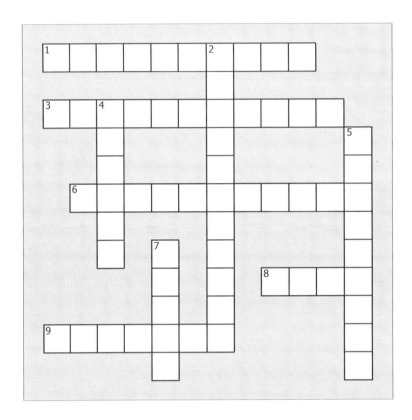

ACROSS

1 The collection of actions and observations performed to verify or falsify a hypothesis.

3 To collect information by using one's senses.

6 Degree of heat or cold.

8 Amount of quantity.

9 A scientist that studies chemistry.

DOWN

2 The comparison of a desired quantity with a standard.

4 The rules that must be obeyed in a laboratory.

5 Study of matter and its changes.

7 A hazard warning symbol for chemicals.

MATTER

INTRODUCTION
TO CHEMISTRY

INTRODUCTION

In chemistry, matter is simply everything; every physical body or substance. Matter has mass and occupies volume.

Matter = Substance(s)

If you remember, we defined chemistry as the study of matter and its changes in the previous chapter.

Look at the picture given below, the mountains, rocks, trees, and lakes pictured are all composed of thousands of substances.

In the physical world, everything can be accepted as matter.

1. MATTER

In this chapter we will study the states of matter, its classification and properties. In addition to these, we will see how different substances can be separated.

In daily life, we frequently ask, "what is the matter?" or "what is the matter with you?". Here, matter means how you are doing or what's going on with you.

2. STATES OF MATTER

Matter can be found in different states, or phases, in the universe. The most common states are: solid, liquid and gaseous (Figure 1). Plasma is often called the fourth state of matter.

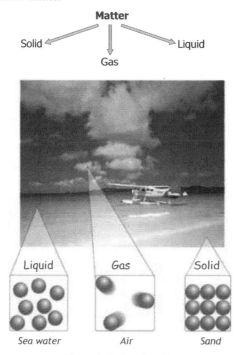

A plasma TV became popular
during the last decade.

Figure 1: States of matter

Through heating and cooling (or changing pressure), matter may alter from one state to another, (names and directions of these changes are given) as follows:

Gaseous (adjective)
Gas (noun)

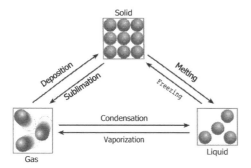

2.1. SOLID STATE

Solid substances have a definite volume and shape. The particles (atoms or molecules) in solids are very close to each other - there is a minute amount of space between atoms or molecules. Solids can be picked up and carried around without a special container.

All pencils and their containers are solid and have a certain shape.

Various solids have different shapes.

2.2. LIQUID STATE

Liquids have definite volume, but no definite shape. Liquids can flow, be poured, and take the shape of their container. The particles in liquids are more loosely contained than those of solids. Hence, that's how liquids can flow.

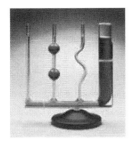

Liquids take the shape of their containers.

Different liquids

2.3. GASEOUS STATE

Gases have no definite volume and no definite shape. A gas takes the shape and fills the volume of any container in which it is placed. Gas particles are apart from each other so they can move freely.

Gases will spread out if they are not in a container. Most gases are colorless, and, therefore, cannot be seen.

Although many gases are colorless, some of them are tinted. For example, nitrogen dioxide (NO_2) has a reddish - brown color and is an extremely toxic gas!

We all live in a gaseous world.

Lightning

Aurora

Plasma
and
Our Lives

Plasma
has no definite volume or shape
and contains electrically charged particles.
Plasmas are collections of freely moving particles.
Plasma temperatures may change, but they are
generally very hot (a few thousands to millions of
degrees °C). Fluorescent light and high - intensity arc
lamps are some examples of where plasma can be seen. In
addition to these, many products today are manufactured
using plasma technologies. Computer chips, aircraft
parts, systems for safe drinking water, high efficiency
lighting products are all examples of these
technologies. The word *Plasma* entered
our dictionaries in 1929.

Stars in galaxies

Flame

3. CLASSIFICATION OF MATTER

Matter exists in millions of different forms in the world. Water is matter just like gold. As ice-cream is composed of different states, so is the sun. Matter can be easily classified according to its purity, as follows:

Particles in elements and compounds cannot be seen with the naked eye.

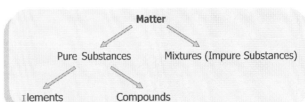

3.1. PURE SUBSTANCES

Tap water (city water) is not a pure substance because it not only contains water molecules, but also particles. Tap water contains other ions, such as calcium, which causes hardness of water.

Pure substances are elements and compounds. They have only one type of particle in their structure. For example, pure water only contains water molecules, and gold solely gold atoms (Figure 2).

Aluminum is an element.

NaCl is a compound.

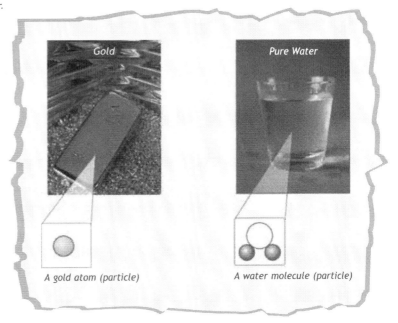

Figure 2: Examples of pure substances

Elements

Elements contain only one type of particle (the atom). All elements are shown by the use of symbols; gold (Au), oxygen (O) and calcium (Ca) are some examples of elements. Today, 116 elements are known; 92 of which are called natural elements. Elements can be classified as metals and nonmetals.

All symbols for elements are placed on a special table called the **periodic table**.

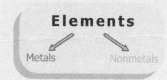

Hydrogen

						110	111	112	113	114	115	116	117	118
Rf	Db	Sg	Bh	Hs	Mt	Ds	Rg	ub		uq		uh	

Table 1: The Periodic Table of Elements is the most popular table in chemistry

He, Ne, Ar, Kr, Xe, and Rn are named Noble gases. These are unreactive and very stable elements (Table 1).

The symbols for elements can have up to two letters in Latin. The first letter is always capitalized and the second must be lower - cased. For example, H : Hydrogen and Al : Aluminum

(If element is unnamed then three letters such as Uub, Uuq etc. temporarily are used in symbols).

Properties of elements.

1. An element cannot be broken down into another substance.

2. The basic building blocks of elements are atoms.

3. When elements react with each other, they produce compounds.

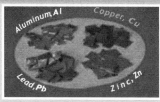

Metals

1. They are good conductors of electricity.
2. They are solid at room conditions.
3. They have a metallic, shiny color.
4. They can be hammered into sheets.
5. They can be drawn into wires.

Nonmetals

1. They do not conduct electricity, except for carbon (graphite).
2. They can be solid, liquid or gaseous at room conditions.
3. They have a dull color.
4. They are brittle (cannot be hammered).
5. They cannot be drawn into wires.

Compounds

Salts, acids, bases, and oxides are all different classes of compounds. Unlike the elements that only amount to 116, there are millions of compounds in the world. All compounds are shown by formulas. For example, H_2O for pure water and CO_2 for carbondioxide. All compounds contain at least two types of particles (atoms).

About 29,000,000 compounds are known.

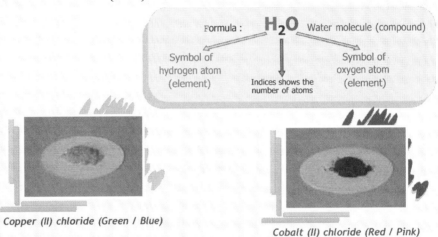

Formula : H_2O Water molecule (compound)

Symbol of hydrogen atom (element)

Indices shows the number of atoms

Symbol of oxygen atom (element)

Copper (II) chloride (Green / Blue)

Cobalt (II) chloride (Red / Pink)

Properties of Compounds

1. A compound can be decomposed into components through chemical methods.
2. Elements combine in definite proportions by mass to form compounds.
3. The chemical properties of compounds are different from those of elements found in that compound.

Copper (II) oxide(Black)

Ammonium dichromate (Orange)

Compounds must be kept in closed containers.

3.2. MIXTURES

Mixtures are combinations of two or more pure substances. In mixtures, the chemical properties of the starting substances do not change. Mixtures can be homogeneous or heteregeneous. In a homogeneous mixture, the composition of its parts are equal. But in heteregeneous mixtures, the composition of its parts is different.

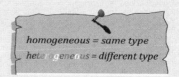

homogeneous = same type
heterogeneous = different type

A heterogeneous mixture
(sand + water + sulfur)

Properties of Mixtures

1. Components of a mixture have their respective chemical properties.
2. Mixtures may be composed of different elements and compounds.
3. There is no fixed ratio among the components.
4. Components can be easily separated by physical means.

A homogeneous mixture

Fizzy drinks are also mixtures.

Unlike compounds, mixtures can be separated into their parts (components) by using physical methods.

Types of mixtures according to physical states

Homogeneous Mixtures (Solutions)		Heteregeneous Mixtures	
State	Examples	State	Examples
Solid	Coins, dental fillings	Solid	Mineral ores
Liquid	Fizzy drinks	Liquid	Milk
Gas	Air	Gas	Aerosols

Homogeneous mixtures are called SOLUTIONS.

Matter	Volume (%)	Mass (%)
Nitrogen	78.09	75.5
Oxygen	20.95	23.15
Argon	0.93	1.29
Carbon dioxide	0.03	0.046
Water vapor and other gases	Very little	Very little

Table 2: Composition of air

In the solar system, only the Earth has the sufficient amount of oxygen for life.

Air is a colorless, odorless and tasteless gas mixture that forms the atmosphere of our Earth. Air is mainly composed of N_2, O_2, Ar, CO_2 and water vapor (Table 2). These gases combine in definite proportions at standard temperature and pressure 22.4 L of dry air weighs 29 g. The air is denser at lower altitudes, whereas density decreases at higher altitudes.

The whole atmosphere is assumed to be 8 km in height and has a uniform thickness.

The oxygen in the air is one of the most important gases for living organisms. At the same time, it is an essential gas for combustion processes. In fact, the quantity and relative percentage of oxygen is so very well defined that everything would be burnt if the percentage of oxygen were 50% instead of 21%.

Air contains different gases.

On the other hand, if the amount of oxygen in air were less than 10%, we would not be able to breathe and would suffocate. On other planets, oxygen is almost non - existent. Instead there is a thick, dark carbondioxide layer and methane gas, both known to be poisonous.

Another important function of carbondioxide in the atmosphere is to block the sun's radiation.

As a result of smog and other gases, there are trace amounts of carbon monoxide, hydrogen sulfide, sulfur dioxide, and other gases in the atmosphere.

These gases evolve from the burning of sulfur-containing fuels from cars and factory chimneys to cause air pollution. On our planet, life can only be possible with clean air. Consequently, we should consider to decrease air pollution.

Air pollution is a problem in many cities.

4. PROPERTIES OF MATTER

Matter has certain properties. Some of them, such as mass and volume, are common for all substances. Other properties like boiling point, density, solubility, etc... are characteristic to each type of substance. Now, let's see how we can classify them:

Every substance has two kinds of properties.

1. Chemical Properties

2. Physical Properties

4.1. CHEMICAL PROPERTIES

Chemical properties are properties that change the nature of matter. Flammability, aciditiy, basicity, and reactivity with water are some examples of chemical properties. When the chemical properties of a substance are altered, it means a chemical change (new substance formed) occurred.

Flammability is a chemical property.

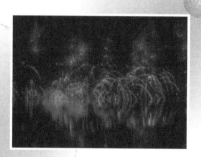

Fireworks are comprised of tiny metals.

The rusting of iron is a chemical change.

4.2. PHYSICAL PROPERTIES

Physical properties are the properties of a substance that can be observed and measured without altering the substance.

Physical properties can be organized as *intensive* and *extensive.*

Extensive Properties

Extensive properties of matter depend on the amount of matter involved. Extensive properties are also called *common properties*, such as *mass, weight, volume, length*, and *charge.*

Mass is an extensive property.

Intensive Properties

Intensive properties matter *do not* depend on the amount of matter given. Intensive properties are sometimes called *distinctive,* or *characteristic*, properties. Color, odor, solubility, hardness, heat/electrical conductivity, melting/freezing point, boiling point, density, luster, ductility, malleability, etc. are all intensive properties.

Luster : Shiny
Ductility : Ability to be bent
Malleability : Ability to be hammered

Mercury is a very toxic (poisonous) substance!

Mercury :
Liquid at room temperature
Melts at –37.9 °C
Boils at 357 °C
Silvery shiny color
Odorless
Good electrical conductor
Has a density 13.6 g/cm³ at 20 °C

Mercury and some of its intensive properties

When a substance's physical properties change, it goes through a physical change.

Some Important Physical Properties

a. Density

Density is the relation between the mass of a substance and its volume. It is denoted by *d* or ρ (rho). If a unit of mass is expressed in g and the unit of volume in cm^3(mL), then the unit of density becomes *g/cm³.*

$$\text{Density} = \frac{\text{mass}}{\text{volume}}$$

$$g/cm^3 \leftarrow \rho = \frac{m \;\to\; g}{V \;\to\; cm^3}$$

The density of a substance is given at constant temperature because its density changes when its volume is altered by a change in temperature. Elements and compounds have characteristic densities at definite conditions (Table 3).

Solid		Liquid		Gas	
Name	Density(g/cm³)	Name	Density(g/cm³)	Name	Density(g/cm³)
Gold	19.3	Water (at 4°C)	1.00	Ammonia	$7.70.10^{-4}$
Aluminum	2.70	Olive oil	0.91	Nitrogen	$1.25.10^{-3}$
Copper	8.92	Gasoline	0.88	Air	$1.29.10^{-3}$
Iron	7.86	Mercury	13.6	Hydrogen	$8.40.10^{-5}$
Silver	10.5	Ethyl alcohol	0.78	Carbondioxide	$1.86.10^{-3}$

Table 3: Densities of some substances at 25 °C and 1 atmospheric pressure

Example 1

Calculate the density of 81 g of an aluminum metal bar with a volume of 30 cm³?

Solution

By using the formula, the density of aluminum can be calculated.

$$\rho = \frac{m}{V} \quad \Rightarrow \quad \rho = \frac{81\,g}{30\,cm^3} = 2.7\,g/cm^3$$

b. Melting Point (mp) and Freezing Point (fp)

Melting point is the temperature at which a solid starts to transform into a liquid. **Freezing point** is the temperature of the reverse change (liquid to solid). In other words, the melting point and freezing point of a pure substance occur at the same temperature. For instance, water melts (freezes) at 0 °C and under 1 atm air pressure.

Element	Melting Point(°C)
Hydrogen	259
Oxygen	218
Aluminum	660
Helium	272
Gold	1063
Iron	1535
Platinum	1769

Melting points of some elements

Ice is frozen water.

Compound	Boiling Point(°C)
Water	100
Naphthalene	217
Ethyl alcohol	78
Butane	1

Boiling points of some substances under 1 atm pressure.

Evaporation may occur at every temperature, but boiling only occurs at the boiling point.

Boiling points depend on pressure. For example, water boils at 100 °C at sea level.

c. Boiling Point (bp)

A boiling point is a temperature at which a liquid transforms into its gaseous state. Actually, a liquid may change to its gaseous state below its boiling point. This process is called evaporation, and it only happens on the liquid's surface. In the boiling process, all molecules in a liquid may be ready to alter to their gaseous state. Boiling happens when the vapor pressure of a liquid becomes equal to air pressure.

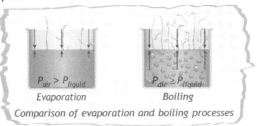

$P_{air} > P_{liquid}$ $P_{air} \geq P_{liquid}$

Evaporation   Boiling

Comparison of evaporation and boiling processes

Boiling point of water at different altitudes*

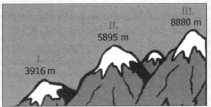

Peak I.   Mt. Erciyes (Turkey)   : Bp of water ~90 °C
Peak II. Mt. Kilimanjaro (Tanzania)   : Bp of water ~80 °C
Peak III. Mt. Everest (China and Nepal)   : Bp of water ~70 °C
*Atmospheric pressure decreases when we go up in the atmosphere.

Example 2

What are the physical states of X,Y and Z at room temperature?

	mp	bp
X :	10	56
Y :	−250	−59
Z :	350	1506

Solution

Substances are liquids at temperatures between their melting and boiling points. Hence, at room temperature X is a liquid, Y is a gas, and Z is a solid.

d. Solubility

Solubility is the amount of substance (solute) dissolved in a given solvent at a given temperature. Dissolving, or dissolution, means the disappearance of a solute in a given solvent. After the dissolving process, solutions are produced. Many chemical compounds can be dissolved in water. Table 4 shows the solubilities of different substances in 100 g of water at 20 °C.

Solute is a substance that dissolves in a solution.

Solvent is a substance that dissolves a solute in a solution.

Solution is a homogeneous mixture of solute and solvent.

Dissolution *is simply the mixing of a solute in a solvent.*

Substance	Solubility (g/100gH_2O)
Table salt	36
Sugar	190
Sodium nitrate	88
Lead (II) nitrate	52
Potassium dichromate	11
Baking soda	10

Table 4: Solubilities of some substances in 100 grams of water at 20 °C.

For example, 11 g of potassium dichromate can only be dissolved in 100 g of water, as shown in Figure 3.

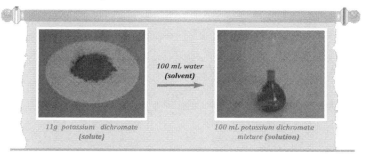

100 mL water
(solvent)

11g potassium dichromate
(solute)

100 mL potassium dichromate
mixture (solution)

Figure 3: Solubility of potassium dichromate is 11g/ 100g water

Even tough many substances dissolve in water, some substances like sulfur cannot dissolve in water (Figure 4).

sulfur
(solute)

water
(solvent)

Figure 4: Sulfur is not water-soluble.

5. SEPARATION OF MIXTURES

As we previously learned, mixtures are not pure substances. In order to obtain one of the components in a mixture, we need to separate them. The separation of mixtures can only be possible when we use the physical properties of substances.

For different types of mixtures, different methods are needed. Now let's see some of these methods used to separate mixtures.

5.1. BY MEANS OF THE USE OF ELECTRICITY

Some mixtures can easily be separated if one component in the mixture is attracted by an electrified object. For example, when an electrified ebony rod comes into contact with a pepper (isot) - salt mixture, the rod attracts small pepper particles, which then become separated from the table salt (Figure 5).

Isot is a type of black-red colored pepper used in Turkey.

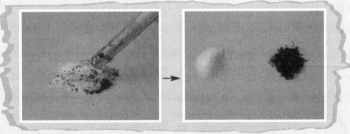

Figure 5: Separation of pepper (isot) and salt mixture by using an electrified ebony rod.

5.2. BY MEANS OF THE USE A MAGNET

Some substances can be separated from a mixture through the use of a magnet. Since iron, nickel and cobalt are attracted by magnets, they can be separated from other substances.

As an example, an iron - sulfur mixture can be separated by using the magnetic property of iron (Figure 6).

Figure 6: Separation of an iron from sulfur in a mixture by using a magnet.

5.3. BY MEANS OF DENSITY DIFFERENCES

If immiscible liquids, which have different densities, are mixed in a container, the denser liquid settles at the bottom and the lighter one at the top. This type of mixture can be separated by using a separatory funnel as shown in Figure 7.

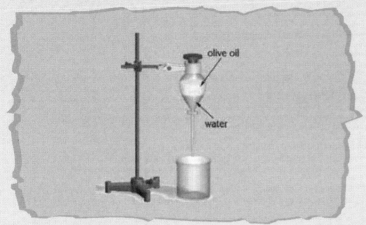

*Liquids which do not mix with each other are called **immiscible** liquids, such as olive oil and water. Liquids which mix with each other in any ratio are called **miscible** liquids such as water and ethyl alcohol.*

Figure 7: *Separation of an olive oil-water mixture*

5.4. BY MEANS OF SOLUBILITY

The solubilities of pure substances are generally different in a solvent. For example, some substances are soluble in water, and others are not. It can be said that the solubility of a substance in water is characteristic under given conditions.

The separation of a copper (II) chloride-sulfur mixture (shown Figure 8) can be achieved by using the solubility differences of the components in water. When this mixture (**8a**) is placed in water, copper (II) chloride will dissolve. Whereas, the sulfur will not (**8b**). If this mixture is filtered, the sulfur particles will be removed through a filter paper (**8c**). The copper (II) chloride solution will then be heated to evaporate the water to obtain copper (II) chloride (**8d**).

Filtration is a method to separate two or more subtances. In filtration, a filter paper is placed in a funnel.

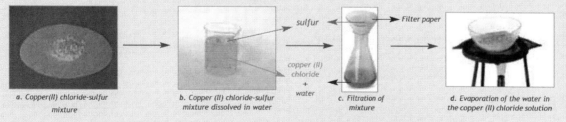

a. Copper(II) chloride-sulfur mixture

b. Copper (II) chloride-sulfur mixture dissolved in water

c. Filtration of mixture

d. Evaporation of the water in the copper (II) chloride solution

Figure 8: *Separation of a copper (II) chloride - sulfur mixture.*

A substance that is nonsoluble in water may be soluble in another liquid.

5.5. BY MEANS OF THE PROCESS OF DISTILLATION

As we previously stated, the melting and boiling points of substances vary. Using these variations, it's possible to separate liquids having different boiling points in a mixture. **Distillation** is a process to separate mixtures by their different boiling points. Now, let's see two types of distillation:

It's possible to separate a mixture of liquids that have different melting points.

1. *Simple distillation* is the separation of a liquid from a solution, such as water from sea water (Figure 9).

2. *Fractional distillation* is a method used to separate a mixture of miscible (mix with each other) liquids (Figure 10). For example, gasoline from petroleum.

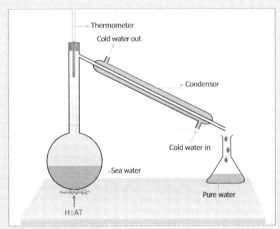

Figure 9: Simple distillation

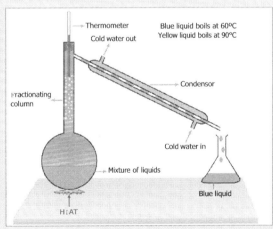

Figure 10: Fractional distillation

PAPER CHROMATOGRAPHY

Paper *chromatography* is a separation method used for mixtures in chemistry. It is used to separate mixtures of color, amino acids, etc. with the help of their solubilities. The mixture sample is absorbed onto *chromatography* paper. The paper is then dipped into a suitable solvent (such as ethanol) and placed in a container. The end result will demonstrate that different compounds in the sample mixture travel various distances, according to how strongly they interact with the paper.

a. Paper and solvent

b. Drops of mixtures

c. Each drop is moved at a different rate.

Through the process of paper chromatography, mixtures can be separated into components.

1. What is the most common type of matter we use in daily life? Discuss in the classroom.

2. Can you cite an example of non-matter?

3. How many states of matter exist? What are the differences between them?

4. Where can we see solid, gaseous and liquid states in the human body?

5. Classify the following as pure substance or mixture.

   a. Bread   e. Orange juice   i. Air

   b. Jam   f. Snow flake   j. Oxygen

   c. Ice (water)   g. Milk   k. Exhaust gas

   d. Soap   h. Sea water   l. Carbon dioxide

6. What is the name of the table that shows all elements? Why is it useful?

7. Why are some elements (gold and platinum) more expensive than others (aluminum and iron)? Research.

8. Look at the periodic table (in Appendix D) and find the name of the following elements.

   a. He   b. U

   c. Ag   d. N

9. Classify the various homogeneous mixtures according to their physical states and give examples for each.

10. What are the components of the following mixtures? Research.

    a. Sea water   c. 18 K Gold

    b. Cough syrup   d. Air

11. How would you measure the real volume of sugar found in a cup (200 mL)? Research. Discuss the results with your friends.

12. How can you measure the density of this book?

13. What is the volume of an iron metal bar that weighs 157g? (ρ_{iron} = 7.86 g/cm^3)

14. What are the methods used in your kitchen by your mother and methods used in a chemistry laboratory to separate mixtures? Compare these methods.

15. What is sieving? Where is it used in daily life? Research.

16. Why do we need to know the melting and the freezing points of substances?

17. If oxygen gas is not soluble in water, what would happen to seawater? Research.

18. Separate the following mixtures by using separation methods, and indicate the minumum amount of steps necessary for separation.

    a. Chalk dust + table salt

    b. Iron powder + water + wood

    c. Alcohol + water + pepper

19. Why is (pure) water not suitable for drinking? Research.

20. Explain the reasons why the labels "Drink cool" or "Store in a cold place" appear on cans/bottles of fizzy drinks? Research.

MULTIPLE CHOICE QUESTIONS

1. Which of the following is not matter?

   A) Chalk   B) Milk   C) Snow

        D) Light   E) Wood

2. Which element below is found in its liquid state at room conditions?

   A) Mercury   B) Aluminum   C) Gold

        D) Oxygen   E) Copper

3. Which one(s) of the following statements is/are correct?

   I. Gases have definite shapes.

   II. All liquids flow at the same speed.

   III. Petroleum (raw oil) is a mixture.

   A) I only   B) II only   C) III only

        D) I and II   E) II and III

4. Which of the following is not a compound?

   A) Table salt   B) Sugar   C) Water

        D) Ammonia   E) Bread

5. What is hard water?

   A) Solid water   B) Difficult water

   C) Water with some ions   D) A type of music

        E) Pure water

6. Which of the following is not a physical property?

   A) Flammability   B) Boiling Point   C) Density

        D) Solubility   E) Conductivity

7. Which of the folliving is an intensive property?

   A) Mass   B) Volume   C) Weight

        D) Solidity   E) Length

8. Which separation method should be used to separate a water - sugar mixture?

   A) Filtration   B) Distillation   C) Sieving

        D) Chromatography   E) Precipation

9. Which one(s) of the following mixture(s) could be separated by fractional distillation?

   I. Salt from salty water

   II. Sugar from a sand - sugar mixture

   III. Gasoline from petroleum

   A) I only   B) II only   C) III only

        D) I and II   E) II and III

10. Sugar dissolves in water, whereas naphtalene does not. In order to separate a sugar - naphtalene mixture to obtain pure sugar, which of the following processes, and, in which sequence, must be followed?

    I. Evaporation

    II. Dissolution in water

    III. Filtration

    A) III, II, I   B) II, III, I   C) I, III, II

         D) III, I, II   E) II, I, III

CRISSCROSS PUZZLE

ACROSS

3 Impure substances

5 A state of substance

7 Mass - volume ratio

9 Common name for elements and compounds

11 Temperature at which solids transform into liquids

12 A type of pure substance

DOWN

1 Temperature at which liquids transform into solids

2 Substance composed of at least two different elements

4 The physical property that depends on an amount of matter.

6 The state of water at room temperature

8 The state of air at room temperature

10 Everything that occupies volume and has mass.

ATOM

INTRODUCTION
TO CHEMISTRY

INTRODUCTION

For ages, humankind has been interested about the structure of substances. For centuries, thinkers, philosophers, alchemists, and scientists have all tried to discover the most fundamental unit of matter. How is matter made up? What are its fundamental substances? How are its structures? In the past, the answers to these questions were not easy to formulate. Let's try a simple exercise in order to better understand just how hard this must have been:

Take a square piece of paper with dimensions of 5 cm · 5 cm. First, tear the paper in half, then take one of the halves and tear again. Continue this process up to 10 times (Figure 1). Hopefully, you were able to do so.

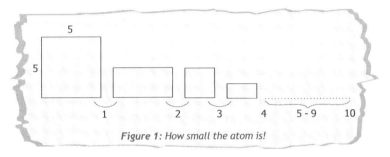

Figure 1: How small the atom is!

If you do this 50 times, how small will that piece of paper become (probably you cannot achieve this)? How can you call this smallest part of paper?

For the last 200 years, scientists have accepted that the smallest parts of substances are called **atoms**.

An atom is a world, according to some scientist, and it works with a mechanism that is not completely understood. But this much is known, an atom has basically two parts:

A **nucleus** (central part) and **electrons** (very fast moving particles) around this center.

A STM

 STM

Q. Is it possible to see an atom?
A. No. But, Scanning Tunneling Microscope (STM) gives us the chance to study and view individual atoms on the surface of materials. STM was invented in 1981 by Gerd Binnig and Heinrich Rohrer in Switzerland. These scientists won the Nobel Prize in Physics (1986).

1. ATOM

Atom is a term that originates from the Greek word ***atomos***, which means indivisible. Atoms are accepted as indivisible because of their minute size. When we discuss atomic size, we use nanometers (0.000000001m scale). For example, the diameter of an atom ranges from about 0.1 to 0.5 nanometer (Figure 2).

1 nanometer (nm) = One billionth of a meter

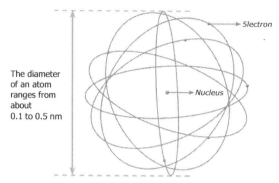

The diameter of an atom ranges from about 0.1 to 0.5 nm

Nucleus

5lectron

Nanotechnology *is the design, study and control of matter using the nanometer scale. This word was first used (defined) in 1974 by Japanese scientist Prof. Norio Taniguchi at Tokyo Science University.*

Figure 2: *Atomic diameter ranges from about 0.1 to 0.5 nm.*

An atom is made up of two main parts, the nucleus and electrons.

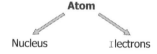

Atom

Nucleus   ɪlectrons

History of The Atom

It's impossible to weigh or isolate a single atom. We know that the atom is so tiny that we cannot feel it by using our senses. Therefore, the development of atomic theories have been based on indirect methods. At the beginning of the 19th century, the existence of atoms had been a point of contention. Erstwhile, Muslim scientist Jabir Bin Hayyan (721-815) had discovered that an individual atom could be split to release a huge amount of energy.

John Dalton (1766-1844) presented the first scientific proof of the existence of the atom based upon his experimental study. At the start of the 20th century, the theories and empirical studies conducted by Thomson, Rutherford, Planck, Einstein, Bohr, Shrodinger... that delved into atom's structure greatly impacted today's progress.

Computer image of chromium on iron atoms.

2. SUBATOMIC PARTICLES

Up to the 20th century, atoms were accepted as indivisible (it cannot be divided into small parts). Today, we know otherwise.

Atomic radius of the hydrogen (H) atom is 37 picometer (37 · 10⁻¹²m)!

During the last century, nuclear reactions showed us how an atom's division is possible. With the help of nuclear reactions, an atom's nucleus can be divided into parts to produce huge amounts of energy. Now let's answer the next question: What are the subatomic particles of an atom? The shape of an atom is assumed to be a simple sphere, but it actually might be an empty sphere (Figure 3).

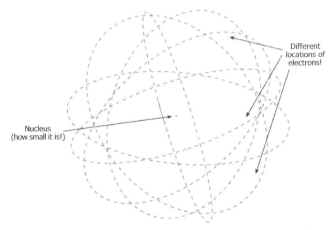

Figure 3: A representation of an atom

In chemistry, the term nucleus (plural nuclei) is used for atoms as it is used for the nucleus of a cell in biology.

In an atom sphere, there is a nucleus at the center. An atom's nucleus contains protons and neutrons. Electrons are also found in an atom, but they move so quickly that we only know that they are around the nucleus but not exactly where?!

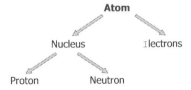

Proton (p) : Positively charged particles found in the nucleus of an atom.

Neutron (n) : Uncharged particles found in the nucleus of an atom.

Electron (e⁻) : Negatively charged particles found around the nucleus of an atom (Table 1).

	Symbol	Charge	amu	Charge(C)	Mass(kg)
Proton	p	+1	1	$1.6 \cdot 10^{-19}$	$1.67 \cdot 10^{-27}$
Neutron	n	0	1	$1.6 \cdot 10^{-19}$	$1.67 \cdot 10^{-27}$
Electron	e	−1	~1/2000	0	$9.11 \cdot 10^{-31}$

Table 1: Subatomic Particles

Amu : atomic mass unit
1 amu means 1/12 of the mass of a carbon - 12 nucleus.

1 Coulomb (C) : Amount of electrical charge carried by a current of 1 ampere (A) in 1 second.

Although an atom was once known as the smallest particle, today we know that even smaller particles exist in an atom (Figure 4).

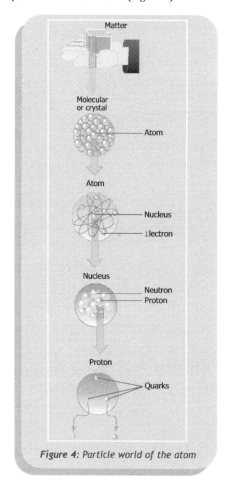

Figure 4: Particle world of the atom

Inside the atom

In the 20th century, we have learned that there are subatomic particles in the atom, such as the proton, neutron and electrons.

But today, scientists relate to the atoms' subparticles such as quarks, leptons, neutrinos, muons, photons, mezons, etc...

The lightest basic particle of an atom, which has a mass of approximately 1/2000 amu, is the electron. For the neutral atom, the number of protons and electrons are equal. Electrons rotate at a great speed at specific, fixed energy levels (shells) around the nucleus. The first shell is the nearest to the nucleus, and is called the "K" shell. The second shell is known as "L", the third is "M" ... etc. Each shell has a certain capacity of electrons, and this capacity is defined by the equation of $2n^2$. Here, n shows the number of shells.

If an atom only has one shell, it can have up to 2 electrons. An atom's electrons in its outermost shell are called **valence electrons**.

Thus, the first energy level (K) can occupy a maximum of 2 electrons, the second energy level (L) can have a maximum capacity of 8 electrons, and so on.

The maximum number of electrons found in each shell can be summarized as such:

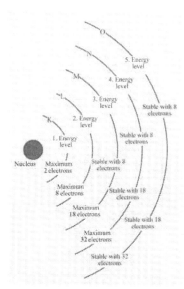

Configuration of electrons in their shell.

First shell (K)	for n = 1	$\Rightarrow$	$2 \cdot n^2 = 2 \cdot 1^2 = 2e^-$
Second shell (L)	for n = 2	$\Rightarrow$	$2 \cdot n^2 = 2 \cdot 2^2 = 8e^-$
Third shell (M)	for n = 3	$\Rightarrow$	$2 \cdot n^2 = 2 \cdot 3^2 = 18e^-$
Fourth shell (N)	for n = 4	$\Rightarrow$	$2 \cdot n^2 = 2 \cdot 4^2 = 32e^-$
Fifth shell (O)	for n = 5	$\Rightarrow$	$2 \cdot n^2 = 2 \cdot 5^2 = 50e^-$

Example 1

Show the electron configuration for the following elements and indicate their number of energy levels.

a. H : 1 electron

b. O : 8 electrons

c. Mg : 12 electrons

d. Ca : 20 electrons

Solution

a. H : 1)　　　　Hydrogen has only one energy level.

b. O : 2) 6)　　Oxygen has two energy levels.

c. Mg : 2) 8) 2)　　Magnesium has three energy levels.

d. Ca : 2) 8) 8) 2)　Calcium has four energy levels.

Exercise 1:

Find the number of valence electrons in,

a. Lithium - 3 electrons

b. Aluminum - 13 electrons

3. ISOTOPES

All of the atoms' nuclei of an element have the same number of protons, but their number of neutrons may be different. For example, there are three types of hydrogen atoms (Figure 5), and they differ only in their number of neutrons.

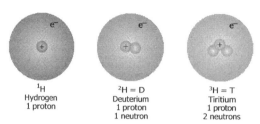

Figure 5: Types of hydrogen atoms

Isotopes are similar to human beings in the world.

Isotopes can be defined as atoms that have the same number of protons but a different number of neutrons (Table 2).

Since the number of protons and electrons of an isotope are equal, isotopes have similar chemical properties but different physical properties.

Element	Atomic number	Number of neutrons	Atomic mass number	Symbol	Natural abundance (%)
Hydrogen	1	–	1	^1H	99.985
	1	1	2	^2H	0.015
	1	2	3	^3H	very small
	3	3	6	^6Li	7.52
	3	4	7	^7Li	92.48
	6	6	12	^{12}C	98.89
	6	7	13	^{13}C	1.11
Oxygen	8	8	16	^{16}O	99.76
	8	9	17	^{17}O	0.04
	8	10	18	^{18}O	0.20
	17	18	35	^{35}Cl	75.4
	17	20	37	^{37}Cl	24.6
Copper	29	34	63	^{63}Cu	69.1
	29	36	65	^{65}	30.9
Uranium	92	143	235	^{235}U	0.71
	92	146	238	^{238}U	99.28

Table 2: The natural abundance of some isotopes

Isotones

Atoms having the same number of neutrons but different number of protons are called **isotones**. For example,

$^{35}_{18}Cl$, Chlorine ⎤
⎥ isotones
$^{36}_{18}Ar$, Argon ⎦

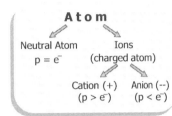

A t o m

Neutral Atom   Ions
$p = e^-$ (charged atom)

Cation (+)   Anion (--)
($p > e^-$)   ($p < e^-$)

Positively charged atoms = CATION

Negatively charged atoms = ANION

Monoatomic ions = Ions with only one atom (Na^+, Cl^-...).

Diatomic ions = Ions with only two atoms (OH^-, NO^-...).

Polyatomic ions = Ions with more than two atoms (SO_4^{-2}, CO_3^{-2}...).

A cation

4. IONS

Atoms of an element are neutral particles. Their number of protons and electrons are equal. Therefore, the charge of a neutral atom is zero.

Otherwise, the atom is charged. Charged atoms are simply called **ions**. There are two types of ions: cations and anions.

If the number of protons is greater than the number of electrons in an atom, it is called a **cation**.

Or, if the number of protons is smaller than the number of electrons in an atom, it is called an **anion**.

The charge of an atom can be determined by the following equation $\boxed{q = p - e^-}$ here,

p : is the number of protons

e^- : is the number of electrons

q : is the charge of an atom

Now let's see some examples of sodium and oxygen atoms;

	Protons	Electrons	Charge	Symbol
Neutral sodium	11	11	$11 - 11 = 0$	Na
Sodium cation	11	10	$11 - 10 = 1$	Na^+
Neutral oxygen	8	8	$8 - 8 = 0$	O
Oxygen anion	8	10	$8 - 10 = -2$	O^{2-}

Example 2

Classify the following atoms as neutral, cation or anion:

a. An oxygen atom (O) has 10 electrons, 8 neutrons and 8 protons.

b. A potassium atom (K) has 18 electrons, 19 protons and 20 neutrons.

Solution

a. For O, p = 8, e^- = 10, n = 8 then $p < e^-$ it is an anion.

   ($q = p - e^- \Rightarrow q = 8 - 10 = -2$)

b. For K, p = 19, e^- = 18, n = 20 then $p > e^-$ it is a cation.

   ($q = p - e^- \Rightarrow q = 19 - 18 = +1$)

Exercise 2:

Find the cations in the following atoms:

a. Fluorine: 9 e^- and 9 p **b.** Gold: 79 p and 78 e^- **c.** Zinc: 30 p and 28 e^-

5. THE ATOMIC TERMINOLOGY

There are some basic terms about the atom in chemistry. In order to study the calculations of the atom, these terms should be understood.

5.1 ATOMIC NUMBER (Z)

The number of protons in an atom is called **atomic number** and is represented by **Z**.

> Atomic number = Number of protons

Each car has a different licence plate number

Each type of atom posesses a different atomic number that specifies its amount of protons. For example, a calcium atom has 20 protons; its atomic number is 20.

*For a **neutral atom**,* the atomic number, the number of protons and the number of electrons are equal.

> Atomic number = Number of protons = Number of electrons

> $Z = p = e^-$

5.2 ATOMIC MASS NUMBER (A)

An atom's total number of protons and neutrons is called its **atomic mass number** and denoted by **A**. An atom's mass is found in its nucleus and can be calculated as follows:

The atomic mass number can sometimes be called nucleon number.

> Atomic Mass Number = Number of protons + Number of neutrons

> $A = p + n$

Example 3

What is the atomic mass number of molibdenum, which has 42 protons and 54 neutrons?

Solution

By using the following formula, molibdenum's atomic mass number can easily be calculated.

$A = p + n$

$A = 42 + 54 \implies A = 96$

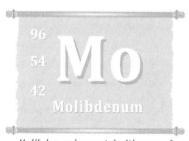

Molibdenum is a metal with a mp of 2610 °C and bp of 5560 °C

Exercise 3:

Calculate the number of neutrons in selenium, (Se) which has 34 protons and an atomic mass number of 79.

Table 3 below shows some basic information about the first 20 elements, from hydrogen to calcium;

Atomic Number (Protons)	Symbol	Name	State at room conditions	Electron in the shells	Discovered
1	H	Hydrogen	gas	2)	1766
2	He	Helium	gas	2)	1895
3	Li	Lithium	solid	2) 1)	1817
4	Be	Beryllium	solid	2) 2)	1798
5	B	Boron	solid	2) 3)	1828
6	C	Carbon	solid	2) 4)	Ancient times
7	N	Nitrogen	gas	2) 5)	1722
8	O	Oxygen	gas	2) 6)	1774
9	F	Fluorine	gas	2) 7)	1884
10	Ne	Neon	gas	2) 8)	1898
11	Na	Sodium	solid	2) 8) 1)	1807
12	Mg	Magnesium	solid	2) 8) 2)	1808
13	Al	Aluminum	solid	2) 8) 3)	1825
14	Si	Silicon	solid	2) 8) 4)	1823
15	P	Phosphorus	solid	2) 8) 5)	1669
16	S	Sulfur	solid	2) 8) 6)	Ancient times
17	Cl	Chlorine	gas	2) 8) 7)	1774
18	Ar	Argon	gas	2) 8) 8)	1894
19	K	Potassium	solid	2) 8) 8) 1)	1807
20	Ca	Calcium	solid	2) 8) 8) 2)	1808

Table 3: The first 20 elements

X Representation

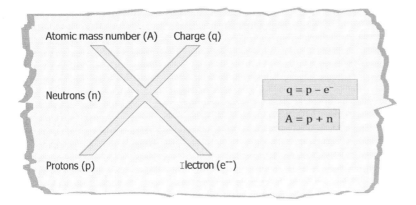

The **X representation** details an atom's number of electrons, number of protons, number of neutrons, charge and atomic mass number.

Here, X represents any symbol of an element such as Al, K, Mg etc... Now, let's see an example. A neutral sodium atom (Na) has 11 protons, 11 electrons, 12 neutrons, and an atomic mass number of 23, which are shown thusly:

$$A = 23 \qquad\qquad 0$$

$$n = 12 \qquad Na$$
$$p = 11 \qquad\qquad e^- = 11$$

Example 4

What is the number of neutrons and the charge of a bromine ion?

$$^{80}_{35}Br^{?}_{36}$$

Solution

If $A = p + n \Rightarrow n = A - p$ and $n = 80 - 35 = 45$

If $q = p - e^- \Rightarrow$ and then $q = 35 - 36 = -1$

Bromine is a very toxic and corrosive liquid.

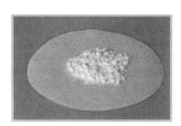

Sulfur

Example 5

If a sulfur ion, S^{2-}, has 16 protons and 16 neutrons. Calculate its;

a) Number of electrons   **b)** Atomic mass number

Solution

a) $q = p - e^-$ $\Rightarrow$ $e^- = p - q = 16 - (-2) = 18$

b) $A = p + n$ $\Rightarrow$ $A = 16 + 16 = 32$

$$^{32}_{16}\text{S}^{2-}_{16} \, _{18}$$

Exercise 4:

What is the number of electrons in a neutral oxygen atom that has 8 neutrons and an atomic mass number of 16?

5.3 RELATIVE ATOMIC MASS AND RELATIVE FORMULA MASS

Elements come together in a certain mass ratio to form compounds. In order to compare these mass ratios, scientists chose one of the atom of the elements as a standard. For this purpose, *^{12}C is accepted as a standard atom.* The mass of the ^{12}C isotope atom is accepted as 12.00 amu and other atomic masses of other elements were calculated accordingly.

For example, hydrogen's atomic mass is equal to 1.008 (= 1 amu) and calcium's atomic mass is equal to 40.078 (= 40 amu) with respect to ^{12}C that is 12 amu.

Relative Formula Mass of H_2O and NH_3 :

$H_2O = (2 \cdot 1) + (1 \cdot 16) = 18\,amu$

$NH_3 = (1 \cdot 14) + (3 \cdot 1) = 17\,amu$

In other words, the relative atomic mass of an element is the average mass of its atom to 1/12 of the mass of a ^{12}C atom. In the periodic table, the masses of atoms are written according to these relative calculations. For compounds; similar to relative atomic mass, a relative formula mass is used. A **relative formula mass** is the sum of relative atomic masses of the atoms found in a compound.

Example 6

Calculate the relative formula mass of CO_2? (Use periodic table in Appendix D)

Solution

Relative formula mass $CO_2 = 1 \cdot$ relative atomic mass C $+ 2 \cdot$ relative atomic mass O

Relative formula mass $CO_2 = (1 \cdot 12) + (2 \cdot 16) = 44$ amu

Exercise 5:

Calculate the relative formula mass of SO_2 and KNO_3?

5.4 AVERAGE ATOMIC MASS

Most of the elements in nature are found as a mixture of isotope atoms. Therefore, determining the atomic mass of these elements can be problematic. For example, the lithium atom has two isotopes; ^{6}Li and ^{7}Li. So, which number will be the atomic mass of Li, 6 or 7? In fact, the atomic mass of Li is exactly 6.94 amu.

Natural abundance of isotopes of lithium	
^{6}Li	7.5 %
^{7}Li	92.5 %

The average atomic mass of lithium is 6.94 ≅ 7 amu

To solve this problem, the average atomic masses are utilized. The **average atomic mass** is the average masses of natural isotopes of an element. The average atomic mass is calculated by multiplying the atomic mass of each isotope by its percentage of abundance and adding the values obtained. This can be shown by the following formula:

Average atomic mass = [(The mass of 1st isotope · the % of abundance of the 1st isotope) + (The mass of 2nd isotope · the % of abundance of the 2nd isotope) +]

Example 7

Chlorine ^{35}Cl and ^{37}Cl isotopes are known. What is the average atomic mass of a chlorine atom, if the percentage of abundance of ^{35}Cl is about 75%, and of ^{37}Cl is about 25%?

 *Chlorine is yellow - green in color and **poisonous** gas.*

Solution

With the help of this formula;

Average atomic mass = [(The mass of 1st isotope · the % abundance of 1st · isotope) + (the mass of 2nd isotope · the % abundance of 2nd isotope) + ...]

Average atomic mass (Cl) = $35 \cdot \dfrac{75}{100} + 37 \cdot \dfrac{25}{100} = 35.5$ amu

Example 8

What is the average atomic mass of Mg, if the natural abundances of Mg isotopes are given below?

^{24}Mg : 78.70%
^{25}Mg : 10.13%
^{23}Mg : 11.17%

Solution

Average atomic mass (Mg) = $\left(24 \cdot \dfrac{78.70}{100}\right) + \left(25 \cdot \dfrac{10.13}{100}\right) + \left(26 \cdot \dfrac{11.17}{100}\right) = 24.3$ amu

A magnesium, Mg, metal ribbon

1. What is the atom? Is it the smallest particle in matter?

2. Compare the properties of protons and neutrons.

3. What is an electron? Explain the difference between valence electrons and other electrons?

4. Write the electron configurations for:

   a. Al_{13}

   b. K_{19}

5. What is an ion? How many types of ions can be found?

6. What is atomic number?

7. What is the definition of isotope?

8. What is the smallest subatomic particle known today? Research and discuss the results in class.

9. Fill the folllowing table with suitable numbers.

Element	p	n	charge	e⁻	A
Mg		12		10	24
Cl		18	−1		35
S	16			16	32
Na	11	12	+1		

10. According to the following table, which atoms are isotopes? (Hint: First fill in the blanks)

Atom	n	p	A
X	70	66	
Y		68	139
Z	73		139
T		70	141

11. What is the X representation? What information about an atom can it detail?

12. Fill in the following blanks.

    a. $_{8}^{...}O_{10}^{2-}$

    b. $_{26}^{...}Fe_{24}^{...}$

    c. $_{20}^{40}Ca_{...}^{2+}$

    d. $_{79}^{197}Au_{76}^{...}$

13. What is the average atomic mass? Compare with relative atomic mass.

14. Element X has 3 isotopes. Abundances of these isotopes are given in the following table.

Isotopes	Abundance (%)
^{20}X	50%
^{21}X	25%
^{19}X	25%

What is the average atomic mass of X?

MULTIPLE CHOICE QUESTIONS

1. Which of the following statements could be a definition of a neutral atom?

   I. An atom having the same number of protons and neutrons.

   II. An atom having the same number of neutrons and electrons.

   III. An atom having the same number of electrons and atomic number.

   A) I only   B) II only   C) III only

        D) I and III   E) II and III

2. Which of the following atoms has a charge of $Z-$?

	P	n	e^-
A) N :	7	7	10
B) Al :	13	14	10
C) Mg :	12	12	10
D) S :	16	16	18
E) P :	15	16	18

3. If Y^{2+} and Y^{2-} are ions of atom Y, which of the following is/are always the same for these ions?

   I. The number of protons

   II. The number of neutrons

   III. The number of electrons

   A) I only   B) II only   C) I and II

        D) II and III   E) I, II and III

4. One Ba^{2+} ion has 54 electrons and its atomic mass number is 137. What is the number of neutrons for Ba?

   A) 80   B) 81   C) 82   D) 83   E) 84

5. Which of the following terms is not related to the atom?

   A) Atomic mass number   B) Atomic number

   C) Average atomic mass   D) Relative atomic mass

        E) Natural mass

6. Which of the following atoms are isotopes?

   $_{16}^{\,?}G_{14}$, $_{15}^{30}H_{?}$, $_{?}^{29}X_{13}$, $_{15}^{28}Y_{?}$, $_{?}^{30}Z_{14}$

   A) X and Y   B) Y and Z   C) Z and G

        D) G and H   E) Z and H

7. How many energy levels (shells) are there in Na_{11}?

   A) 1   B) 2   C) 3   D) 4   E) 5

8. What is the number of valence electrons in Ca_{20}?

   A) 1   B) 2   C) 3   D) 4   E) 5

9. Which of the following atoms has a different electron configuration?

   A) $_{19}K^{1+}$ B) $_{20}Ca^{2+}$ C) $_{18}Ar$ D) $_{16}S^{2+}$ E) $_{17}Cl^{1-}$

10. What is amu?

    A) Atomic master unit   B) Atomic microbes unit

    C) Atomic unit   D) Atomic mass unit

         E) Atom massive unit

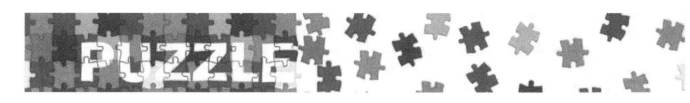

CRISSCROSS PUZZLE

ACROSS

2 Charged atoms

3 Subatomic particles around the nucleus of an atom

8 Basic unit of all substances

9 Total number of protons and neutrons in an atom

11 Negatively charged ion

DOWN

1 Number of protons in an atom

4 Center of an atom

5 A neutral particle in an atom

6 A positively charged particle found in the nucleus of an atom

7 Positively charged ion

10 Atomic mass unit

COMPOUNDS

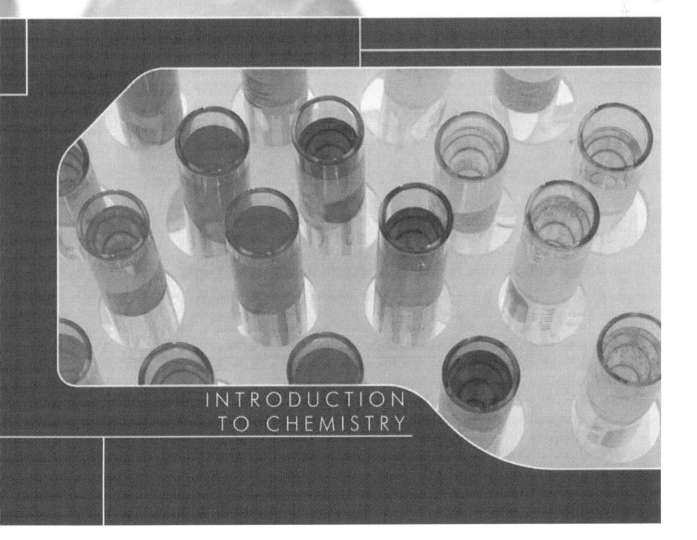

INTRODUCTION
TO CHEMISTRY

INTRODUCTION

In this chapter, we will focus on compounds, their structures and how to name them. As you remember from the previous chapter, matter is classified as pure substances and mixtures. Elements and compounds are pure substances.

Water and sugar are some of the most common compounds in our daily lives.

They are shown by the formulas of H_2O and $C_{12}H_{22}O_{11}$, respectively.

In road construction, many chemical compounds must be used.

The combination of elements in certain ratios result in compounds. In other words, compounds are substances produced as a result of chemical reactions of elements.

EXTRA

The 5 Most Abundant Compounds in the Earth's Crust

About 29 million of compounds are known today

Compound	Formula	Abundance %
Silicon dioxide	SiO_2	42.86 %
Magnesium oxide	MgO	35.07 %
Iron (II) oxide	FeO	8.97 %
Aluminum oxide	Al_2O_3	6.99 %
Calcium oxide	CaO	4.37 %

Compounds can be classified in different ways; here we will classify them as ionic or molecular, according to their bond structures.

Compounds

Ionic   Covalent

1. IONIC COMPOUNDS

Ionic compounds are the compounds which are formed as a result of ionic bonds. Thus, we must first learn the meaning of ionic bond.

Ionic Bond

When an atom donates electrons, a cation is formed. In contrast, when an atom accepts electrons, an anion is formed. In general, metal atoms donate electrons and nonmetal atoms accept electrons easily. For example, a sodium atom (Na) gives an electron and forms one Na^+ cation. On the other hand, a chlorine atom accepts the electron of the sodium atom and forms a Cl^- anion.

An ionic bond holds NaCl in a crystalline structure.

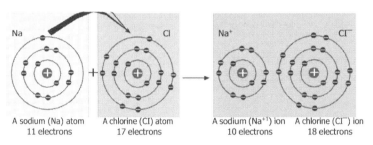

| A sodium (Na) atom | A chlorine (Cl) atom | A sodium (Na^{+1}) ion | A chlorine (Cl^-) ion |
| 11 electrons | 17 electrons | 10 electrons | 18 electrons |

Figure 1: *Formation of sodium and chloride ions*

The electrostatic attraction between cation and anion atoms causes them to bind together. This is called an **ionic bond**. For example, Na^+ and Cl^- attract each other to form NaCl. Since the electrostatic attraction occurs in all directions around a cation or an anion, a cation combines with more than one anion and an anion combines with more than one cation. Thus a huge structure, which is called a crystal, is formed. For example, many Na^+ and Cl^- ions combine with each other to form NaCl crystal.

Briefly, an ionic bond is the bond between metal and nonmetal atoms as a result of an electron transfer.

NaCl crystal seen under a microscope

Properties of Ionic Compounds

1. They are composed of metals and nonmetals.

2. They have ionic bonds.

3. They do not conduct electricity in their solid state. Their aqueous (in water) solutions conduct electricity.

4. They generally dissolve in water and produce ions.

5. They are solid at room conditions.

NaCl is a very popular ionic compound; it is also known as table salt.

2. MOLECULAR COMPOUNDS (COVALENT COMPOUNDS)

The compounds formed with covalent bonds are called molecular compounds. Carbon dioxide (CO_2), water (H_2O), sulfuric acid (H_2SO_4) are some examples of well - known molecular compounds. The smallest unit of molecular compounds are called ***molecules.***

A CO₂ molecule

A H₂O molecule

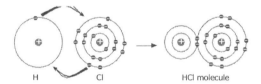

A H₂SO₄ molecule

Firefighters use CO₂ as fire extinguishers.

Most of the compounds that we know today are molecular compounds.

Covalent Bond

The bond between two nonmetal atoms as a result of electron sharing is called a ***covalent bond***. For example, H and Cl atoms form a covalent bond.

H   Cl   HCl molecule

There are two types of covalent bonds.

a. Nonpolar Covalent Bond

Covalent bonds found between molecules composed of the same atoms are ***nonpolar covalent bonds***. For example, F_2 and O_2 molecules have nonpolar covalent bonds.

O_2 F_2

b. Polar Covalent Bond

Covalent bonds formed between atoms with different nonmetals are called ***polar covalent bonds***.

H – Cl

I – Cl

Properties of Molecular Compounds

1. They are composed of nonmetal atoms.

2. They have covalent bonds.

3. They generally do not conduct electricity.

4. They can be in solid, liquid or gas form at room temperature.

3. NAMES OF IONS

In order to write a chemical formula for compounds, the names of their ions should be known. The oxidation number of some important metals and nonmetals is given in Table 1.

a. Monoatomic Cations (Metal ions)

1+	2+	3+
H^{1+} Hydrogen	Mg^{2+} Magnesium	Al^{3+} Aluminum
Na^{1+} Sodium	Fe^{2+} Iron (II)	Fe^{3+} Iron (III)
K^{1+} Potassium	Cr^{2+} Chromium(II)	Cr^{3+} Chromium(III)
Hg^{1+} Mercury(I)	Hg^{2+} Mercury(II)	
Ag^{1+} Silver	Ca^{2+} Calcium	
Cu^{1+} Copper(I)	Cu^{2+} Copper(II)	4+
Li^{1+} Lithium	Pb^{2+} Lead(II)	Pb^{4+} Lead(IV)
	Sn^{2+} Tin (II)	Sn^{4+} Tin(IV)
	Ba^{2+} Barium	
	Ni^{2+} Nickel	
	Zn^{2+} Zinc	

b. Monoatomic Anions (Nonmetal ions)

1−	2−	3−
F^{1-} Fluoride	O^{2-} Oxide	N^{3-} Nitride
Cl^{1-} Chloride	S^{2-} Sulfide	P^{3-} Phosphide
Br^{1-} Bromide		
I^{1-} Iodide		
H^{1-} Hydride		

Monoatomic means "containing only one atom".

The numerical values of charges in atoms are also sometimes called valencies.

The **valency** of elements shows the tendency of atoms to donate or accept electrons in order to stablize their last electron level.

Table 1: The oxidation number of some important metals and nonmetals

Ions that contain more than one type of atom are called **polyatomic ions**. These ions are formed by a combination of atoms. The most important polyatomic ions are shown in Table 2.

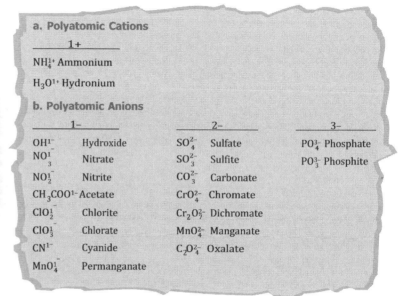

a. Polyatomic Cations

1+
NH_4^{1+} Ammonium
H_3O^{1+} Hydronium

b. Polyatomic Anions

1–		2–		3–	
OH^{1-}	Hydroxide	SO_4^{2-}	Sulfate	PO_4^{3-}	Phosphate
NO_3^{1-}	Nitrate	SO_3^{2-}	Sulfite	PO_3^{3-}	Phosphite
NO_2^{1-}	Nitrite	CO_3^{2-}	Carbonate		
CH_3COO^{1-}	Acetate	CrO_4^{2-}	Chromate		
ClO_2^{1-}	Chlorite	$Cr_2O_7^{2-}$	Dichromate		
ClO_3^{1-}	Chlorate	MnO_4^{2-}	Manganate		
CN^{1-}	Cyanide	$C_2O_4^{2-}$	Oxalate		
MnO_4^{1-}	Permanganate				

Sulfate and sulfite both have 2-valencies.

Table 2: The oxidation number of some important polyatomic ions

Formula of Compounds

In chemistry, as we learned in previous chapters, elements and their atoms are shown by symbols. But compounds are shown by formulas.

A *formula* is a combination of symbols and numbers that represents compounds. Let's see the compound of H_2SO_4 (sulfuric acid).

Subscript 1 is not written in the formulas of compounds.

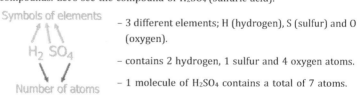

Symbols of elements

$H_2 SO_4$

Number of atoms

– 3 different elements; H (hydrogen), S (sulfur) and O (oxygen).

– contains 2 hydrogen, 1 sulfur and 4 oxygen atoms.

– 1 molecule of H_2SO_4 contains a total of 7 atoms.

Similar to H_2SO_4, each compound has a formula:

H_2O (water) = 2 hydrogen and 1 oxygen atoms

NH_3 (ammonia) = 1 nitrogen and 3 hydrogen atoms

$C_{12}H_{22}O_{11}$ (sugar) = 12 carbon, 22 hydrogen and 11 oxygen atoms

Types of Formulas

1. Empirical Formula is a formula that shows the type and ratio of elements in a compound. ($NO_2 - N_2O_4$)

  N_nO_{2n}
– Nitrogen and oxygen elements
– Ratio between N/O atoms $n/2n = 1/2$

2. Molecular Formula is a formula that shows the actual number of atoms in a compound.

CO_2
– 1 carbon atom
– 2 oxygen atoms.

3. Structural Formula is the formula that shows the way atoms bond, the types of bonds they form and their molecular geometry in space.

– Two O – H bonds
– Angular molecule

Writing Formulas of Ionic Compounds

In order to write a formula for a compound, for example formed by X^{n+} and Y^{m-}, the absolute values of the charges should be crossed, as shown below. Here, **crossing** means writing the number of the charge of an ion as a subscript on the other one.

The sum of the oxidation numbers (valencies) of all elements in a compound is zero.

Example 1

Show how formulas are written for the compounds below:

a. NaCl   **b.** KNO_3

Solution

a. $Na^{+1} \bowtie Cl^{-1}$
Na_1Cl_1 or NaCl

b. $K^{+1} \bowtie NO_3^{-1}$
$K_1(NO_3)_1$ or KNO_3

CO_2, carbon dioxide, has one carbon atom and two oxygen atoms in its formula. Such compounds do not require the subscript 1 to specify their single carbon atom.

Exercise 1:

Write formulas for the compounds formed by the following pair of ions:

a. Na^+ and O^{2-}　　**b.** Ca^{2+} and S^{2-}　　**c.** Al^{3+} and F^-

Finding the Oxidation Number of an Element in a Compound

Since the sum of the oxidation numbers for all of the elements in a compound is zero, we can find the oxidation number of an element in a compound. But you should remember the oxidation number of some common ions like $Na^{1+}, K^{1+},$ $Ca^{2+}, Ba^{2+}, Zn^{2+}, Ag^{1+}, Al^{3+}$ are constant. Additionally, an in general, oxygen has an oxidation number of −2, and hydrogen has +1, in their respective compounds.

Let's examine the examples below.

⟨**Example**⟩ ──────────────────────────── **2**

KNO₃, also known as saltpeter, is used in the manufacture of fireworks and matches; production of fertilizers, and as a food preservative for meat products.

What is the oxidation number (valency) of N in KNO_3?

⟨**Solution**⟩

The sum of the oxidation number of atoms should be zero.

In KNO_3, the oxidation numbers of K and O are 1+ and 2−, respectively.

$$K^{1+} \overset{x}{N} O_3^{2-}$$

$$K + N + 3 \cdot (O) = 0$$
$$(+1) + x + [(3 \cdot (-2)] = 0$$
$$x = +5$$

⟨**Example**⟩ ──────────────────────────── **3**

What is the oxidation number of Cr in $Na_2Cr_2O_7$?

⟨**Solution**⟩

We know that the oxidation numbers of sodium and oxygen are generally +1 and −2, respectively, in their compounds.

$$Na_2^{1+} \overset{x}{Cr_2} O_7^{2-}$$

$$2(Na) + 2(Cr) + [7 \cdot (O)] = 0$$
$$[2 \cdot (+1)] + 2 \cdot x + [7 \cdot (-2)] = 0$$
$$x = +6$$

What is the oxidation number of S in the compound $Al_2(SO_4)_3$?

Solution

Remember that the oxidation numbers of aluminum and oxygen are +3 and –2, respectively, in their compounds.

$$Al^{3+}_2 \ (\overset{x}{SO_4}{}^{2-})_3$$

$$2(Al) + 3[S+4\cdot(O)] = 0$$
$$[2\cdot(+3)] + 3\,[x+4\cdot(-2)] = 0$$
$$x = +6$$

Exercise 2:

What is the oxidation number of lead (Pb) in PbO_2?

What are the Oxidation Numbers of Iron Atoms in Fe_3O_4?

Since the net charge of a compound is zero, the oxidation number of Fe in Fe_3O_4 is calculated below.

$$3\cdot(Fe)+4\cdot(O)=0 \Rightarrow 3\cdot x+4\cdot(-2)=0 \Rightarrow x=+8/3$$

Such a number (8/3) is impossible as an oxidation number; it should be an integer number, because it shows the number of electrons given or taken. Thus, 8/3 as an answer cannot be used. The reason is that Fe_3O_4 is a compound (mixed oxide) that contains both **FeO** and **Fe_2O_3**.

$$Fe_3O_4 = (2 \cdot FeO) + (1 \cdot Fe_2O_3)$$

The oxidation numbers of oxygen and iron are 2– and 2+, respectively, in FeO. Similarly, the oxidation number of oxygen is 2–. Then, the oxidation number of Fe in Fe_2O_3 is;

$$[2 \cdot (Fe)] + [3 \cdot (O)] = 0 \Rightarrow [2 \cdot (Fe)] + [3 \cdot (-2)] = 0, \ Fe = +3$$

Consequently, oxidation numbers of one Fe^{2+} ion and two Fe^{3+} ions are a total of +8 ($1 \cdot 2 + 2 \cdot 3$), and there are four O^{2-}, with a total charge of –8 ($4 \cdot -2$) in Fe_3O_4. Thus, the net charge of Fe_3O_4 is zero.

4. NAMING COMPOUNDS

4.1. NAMING IONIC COMPOUNDS

a. In naming metal-nonmetal compounds, the name of a metal is always written first. Then, the name of the nonmetal ion is added.

Name of Metal + Name of Nonmetal ion

KCl

Potassium   +   chloride

NaCl	Sodium chloride	Na_2S	Sodium sulfide
$CaBr_2$	Calcium bromide	CaO	Calcium oxide
MgI_2	Magnesium iodide	CaC_2	Calcium carbide

C^{4+} : Carbide ion

Example 5

Write and name the formulas of the compounds between the following pairs of elements.

a. Al and S   **b.** Ag and Cl   **c.** Ba and O

Solution

a. Al_2S_3 : Aluminum sulfide

b. AgCl : Silver chloride

c. BaO : Barium oxide

Exercise 3:

What are the formulas of aluminum chloride and potasium iodide?

b. In naming metal and polyatomic anion compounds: first, the name of the metal then, the name of the polyatomic anion are written.

Name of Metal + Name of Polyatomic ion

$CaSO_4$

Calcium   +   sulfate

NaOH	Sodium hydroxide	$AlPO_3$	Aluminum phosphite
$MgSO_4$	Magnesium sulfate	KNO_3	Potassium nitrate
$CaCO_3$	Calcium carbonate	Na_2CrO_4	Sodium chromate

Exercise 4:

Name the following compounds:
a. $KMnO_4$ **b.** $Al(OH)_3$

Exercise 5:

Write and name the formulas of the compounds formed between Al and the following polyatomic ions:
a. SO_4^{2-} **b.** CO_3^{2-} **c.** PO_4^{3-}

c. Some metals, such as Fe, Mn, Cu... etc. may have more than one oxidation number in their compounds. For example, Fe^{2+}, Fe^{3+}, Cu^+, Cu^{2+} etc... In naming metal compounds with variable oxidation numbers (valencies), write the symbol of the metal and indicate the oxidation numbers of the metal in parentheses with Roman numbers. Finally, write the name of the nonmetal ion.

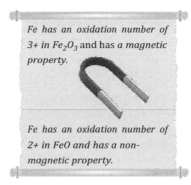

Fe has an oxidation number of 3+ in Fe_2O_3 and has a magnetic property.

Fe has an oxidation number of 2+ in FeO and has a non-magnetic property.

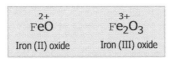

$\overset{2+}{Fe}O$	$\overset{3+}{Fe_2}O_3$
Iron (II) oxide	Iron (III) oxide

A few other examples are given below:

Cu_2O : Copper (I) oxide   MnO : Manganese (II) oxide
CuO : Copper (II) oxide MnO_2 : Manganese (IV) oxide

Exercise 6:

Name the following compounds:

a. $PbCl_4$ **b.** $PbCl_2$

Naming Hydrates
(Water - Containing Ionic Compounds)

In naming hydrates, we have to indicate the number of H_2O molecules by using Greek prefixes and add **hydrate** as a suffix.

Formula	Name	Common Name
$MgSO_4 \cdot 7H_2O$	Magnesium sulfate heptahydrate	Epsom Salt
$CaSO_4 \cdot 2H_2O$	Calcium sulfate dihydrate	Gypsum
$Na_2CO_3 \cdot 10H_2O$	Sodium carbonate decahydrate	Washing soda

4.2. NAMING MOLECULAR COMPOUNDS

In naming molecular compounds, an element's name and its number of atoms are written in Greek, as shown below.

Number +	Name of nonmetal	+ Number +	Name of nonmetal ion

Greek Numbers

Mono	:	One
Di	:	Two
Tri	:	Three
Tetra	:	Four
Penta	:	Five
Hexa	:	Six
Hepta	:	Seven
Octa	:	Eight
Nona	:	Nine
Deca	:	Ten

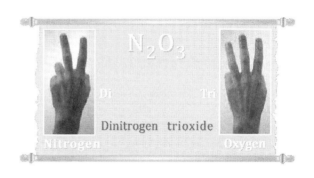

Dinitrogen trioxide

If the number of the first element is one, there is no need to label it as mono. For example, in naming carbon monoxide (CO) we do not need to name it *monocarbon monoxide*.

Some examples of names of molecular compounds are given below.

Note that:
monoxide = monooxide
tetroxide = tetraoxide
pentoxide = pentaoxide

Formula	Name	Formula	Name
SO_2	Sulfur dioxide	PCl_5	Phosphorus pentachloride
PCl_3	Phosphorus trichloride	CCl_4	Carbon tetrachloride
NO	Nitrogen monoxide	P_2O_5	Diphosphorus pentoxide

Some molecular compounds are known by their common names. For example, H_2O is called water, and NH_3 is known as ammonia.

Exercise 7:

Name the following compounds.

a. N_2O_4 **b.** CO_2 **c.** SF_6

d. N_2O_5 **e.** P_2O_3 **f.** CS_2

Law of Definite Proportion (Proust's Law)

Law of Definite Proportion is one of the basic chemical laws. It states that:

Elements of a compound are combined in definite proportion by mass. For example, the mass ratio between hydrogen and oxygen in water is 1/8. This means that 1 amu of H always combines with 8 amu of O atom to form H_2O molecule.

$$H_2O: \frac{m_H}{m_O} = \frac{2}{16} = \frac{1}{8}$$

$$CO_2: \frac{m_C}{m_O} = \frac{12}{32} = \frac{3}{8}$$

$$N_2O_4: \frac{m_N}{m_O} = \frac{28}{256} = \frac{7}{64}$$

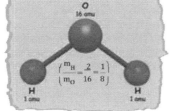

$$\left(\frac{m_H}{m_O} = \frac{2}{16} = \frac{1}{8} \right)$$

Figure 2: A water molecule is 18 amu

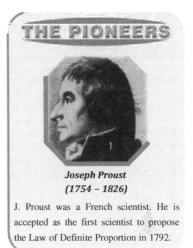

Example 6

The mass ratio $\frac{m_X}{m_O}$ in a compound is $\frac{1}{2}$. How many grams of X and O should be used in order to produce 15 g of compound?

Solution

According to the given ratio, 1 g X and 2 g O form a 3 g compound. Then, $\frac{15}{3}$ = 5 times greater. So, the needed mass of O and X are

$$m_X = 5 \cdot 1 = 5 \text{ g} \qquad m_O = 5 \cdot 2 = 10 \text{ g}$$

Law of Definite Proportion is also known as the "Law of Definite Composition" or the "Law of Constant Composition".

Exercise 8:

What is the mass ratio of elements in SO_3? (S : 32, O : 16)

Law of Multiple Proportion

The law of multiple proportion proposed by J. Dalton in 1803 briefly states that when two elements form more than one compound between them, the ratios of the masses of the second element that combine with the constant mass of the first element will be a positive integer. For example, mass ratios of N are given below.

Compound	m_N	m_O	Ratio
N_2O_4	28	64	$\frac{64}{16} = 4$
N_2O	28	16	

1. What is a compound?

2. What are the differences between elements and compounds?

3. Which compounds are always found in your kitchen?

4. Compare sea water and pure water. Explain the differences between mixtures and compounds.

5. Write the formulas for the following compounds.

   a. Calcium oxide
   b. Aluminum sulfide
   c. Magnesium fluoride
   d. Iron (II) chloride
   e. Tin (IV) oxide

6. Name the following compounds.

   a. AgCl
   b. NaBr
   c. Cu_2O
   d. ZnS
   e. HCl

7. Write the formulas for the following compounds.

   a. Potassium chlorate
   b. Aluminum sulfate
   c. Hydrogen cyanide
   d. Silver nitrate
   e. Zinc hydroxide

8. Name the following compounds.

   a. $Mg(OH)_2$
   b. Na_2CrO_4
   c. $Ca(NO_3)_2$
   d. $FeSO_4$
   e. $(NH_4)_3PO_4$

9. Write formulas for the compounds below.

   a. Dinitrogen monoxide
   b. Oxygen difluoride
   c. Xenon tetrafluoride
   d. Phosphous pentachloride

10. Name these compounds.

   a. NO_2
   b. IF_5
   c. ClF_3
   d. SF_4

11. Some compounds are so expensive. 100 g of silver nitrate ($AgNO_3$), for example, costs $120. Conversely, 100 g of table salt (NaCl) costs $0.1. Why? Research.

12. How many compounds are known in the world? Research.

13. What are the most common compounds found in the atmosphere? Research.

14. Which compounds are mainly found in the human body? Research.

1. Which of the following statements is/are correct for ionic compounds?

   I. Most of them are solid at room temperature.

   II. They conduct electricity in their solid state.

   III. They generally dissolve in water.

   A) I only   B) I and II   C) I and III

        D) II and III   E) I, II and III

2. Which of the following is an ionic compound?

   A) CO   B) KI   C) SF_4 D) PCl_3 E) OF_2

3. In which of the following compounds does chlorine (Cl) have the highest oxidation number? (Hint : O^{2-} and H^{1+})

   A) $HClO_4$ B) $HClO_3$ C) $HClO_2$ D) HClO   E) HCl

4. Which of the following compounds is named incorrectly?

   A) $MgCl_2$: Magnesium chlorine

   B) $Al_2(SO_3)_3$: Aluminum sulfite

   C) CCl_4 : Carbon tetrachloride

   D) N_2O : Dinitrogen monoxide

   E) $CuCl_2$: Copper (II) chloride

5. Which of the following is not a compound?

   A) CuO   B) CO_2 C) Co   D) CaO   E) Cr_2O_3

6. Which of the following does not have covalent bonds?

   A) CO_2 B) SO_3 C) CCl_4 D) KCl   E) NO_2

7. How many atoms are found in $Mg(NO_3)_2$?

   A) 3   B) 5   C) 7   D) 9   E) 11

8. Which of the compounds below is composed of only two atoms? (Hint: X shows only an atom)

   A) X chlorate

   B) X carbonate

   C) X chlorite

   D) X chloride

   E) X chromate

9. Silicon dioxide is known as the most abundant compound in the Earth's crust. According to this information;

   I. It has two different elements.

   II. It is shown as SiO_2.

   III. It is not an ionic compound.

   which of the statement(s) above is/are correct?

   A) I only   B) II only   C) I and II

        D) II and III   E) I, II and III

10. Which one of the following compounds is most prevalent in sea water?

    A) KCl   B) NaCl   C) $MgCl_2$ D) $AlCl_3$ E) $CaCl_2$

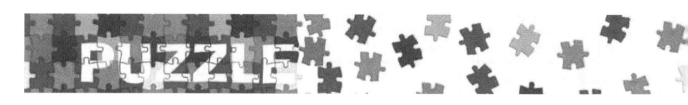

CRISSCROSS PUZZLE

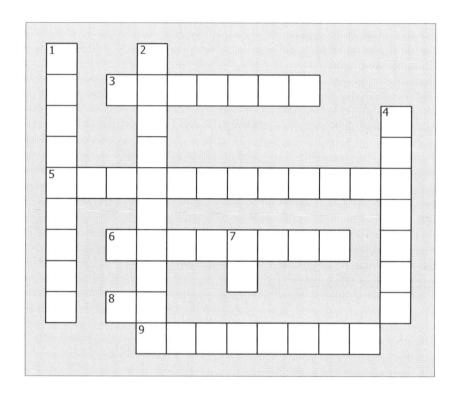

ACROSS

3 Symbols for compounds

5 Bond between nonmetal atoms

6 Covalent bond between the same nonmetal atoms

8 The formula for potassium iodide

9 A type of pure substance

DOWN

1 Bond between metal and nonmetal atoms

2 Ions that contain more than one type of atom

4 Water molecule that contains ionic compounds

7 Symbol for a hydroxide ion

MOLE CONCEPT

INTRODUCTION

In daily life, the use of suitable measurement units enables us to evaluate quantities more precisely. Instead of requesting 123 cherries from your grocer, isn't it easier to ask him for a desired quantity in kilograms? Or, instead of packets of sugar, would you rather ask for 240 lumps in a supermarket (Figure 1)?

123 or 1 kg   240 or a packet

Figure 1: The use of suitable measurement units enables us to better express quantities

*The "mole" is abbreviated as **mol**.*

Similarly, when we study small particles in chemistry we use a special unit known as the mole (mol).

1. MOLE

One of the most important unit in chemistry is the mole. One ***mole*** is accepted as $6.02 \cdot 10^{23}$ particles. Here, the particles may be any atom, molecule, pencils etc...

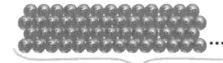

•••••• = 1 mol of iron atoms

602,000,000,000,000,000,000,000 iron atoms

•••••• = 1 mol of people

602,000,000,000,000,000,000,000 people

•••••• = 1 mol of apples

602,000,000,000,000,000,000,000 apples

2. AVOGADRO'S NUMBER

602,000,000,000,000,000,000,000 is a very large number, known as **Avogadro's number**. It is abbreviated as N_A.

Why do we use such huge numbers in chemistry? The answer is quite easy: Because quantities in chemistry can sometimes be very, very small.

For example, 1 hydrogen atom weighs about,

$$1 \text{ amu} = 0.00000000000000000000000166 \text{ g}$$

How is it possible to weigh this amount in a laboratory?

Avogadro's number, $6.02 \cdot 10^{23}$, has proved to be a very good conversion factor. It makes our calculations easier.

$$1 \text{ gram} = 6.02 \cdot 10^{23} \cdot \text{amu}$$

or

$$1 \text{ g} = 1 \text{ mol} \cdot \text{amu}$$

Let's remember again the example of the hydrogen atom:

1 hydrogen atom is 1 amu.

$\underbrace{6.02 \cdot 10^{23} \text{ of hydrogen atoms}}_{1 \text{ mol}}$ is 1 gram.

or 1 mol of hydrogen atoms is 1 gram, it's that simple.

Similar to hydrogen, this conversion can be used for all atoms or molecules. Let's look at the following examples,

I. The atomic mass of a carbon atom is 12 amu, or

1 mol of carbon atoms is 12 g.

II. The atomic mass of a calcium atom is 40 amu, or

1 mol of calcium atoms is 40 g.

III. The atomic mass of a platinum atom is 195 amu, or

1 mol of platinum atoms is 195 g.

Avogadro's number is also known as Avogadro's constant, which refers to $6.02 \cdot 10^{23}$ particles / mole.

1 amu $\cdot$ 6.02 $\cdot$ 10^{23} = 1 g

THE PIONEERS

Amadeo Avogadro(1776–1856)

Avogadro was an Italian lawyer and scientist, who studied physics and mathematics.

He found that equal volumes of gases contain equal numbers of particles under the same conditions.

The atomic mass of each atom is given in the periodic table.

3. MOLAR MASS (M)

A formula

Na - Cl - Na

Cl - Na - Cl

Na - Cl - Na

Because of its crystalline structure, an ionic compound is expressed as a formula, not in molecules like molecular compounds.

The mass of 1 mol ($6.02 \cdot 10^{23}$) of substance in grams is defined as its **molar mass**.

1 mol of sodium atoms (Na)	= $6.02 \cdot 10^{23}$ Na atoms	= **23 g**
1 mol of zinc atoms (Zn)	= $6.02 \cdot 10^{23}$ O atoms	= **65.4 g**
1 mol of oxygen molecules (O_2)	= $6.02 \cdot 10^{23}$ O_2 molecules	= **32 g**
1 mol of carbon dioxide molecules (CO_2)	= $6.02 \cdot 10^{23}$ CO_2 molecules	= **44 g**
1 mol of sodium chloride (NaCl)	= $6.02 \cdot 10^{23}$ NaCl formula	= **58.5 g**
1 mol of hydrogen ions (H^+)	= $6.02 \cdot 10^{23}$ H^+ ions	= **1 g**

1 mol Zn (metal pieces)   *1 mol NaCl (granules)*

Example 1

Calculate the molar mass of water, H_2O.

Solution

The formula for water, H_2O, is composed of 2 hydrogen atoms and 1 oxygen atom. By looking at the periodic table (Appendix D); you can find the mass of a hydrogen atom as 1 amu, and the mass of an oxygen atom as 16 amu.

Molar Mass of H_2O = 2 · (mass of the H atom) + 1 · (mass of O atom)

Molar Mass of H_2O = $(2 \cdot 1) + (1 \cdot 16)$

Molar Mass of H_2O = 18 amu

If a water molecule is 18 amu, then 1 mole of water molecule becomes 18 g.

Remember

1 amu · 1 mol = 1g

Exercise 1:

What is the molar mass of $Al_2(SO_4)_3$?

4. MOLAR VOLUME OF GASES

The volume of 1 mol of gas is called the **molar volume** of that gas. At a standard temperature and pressure (STP), one mole of any gas occupies 22.4 L volume.

STP = 0 ºC and 1 atm pressure

For example:

1 mole of O_2 has   22.4 L volume at STP

1 mole of He has   22.4 L volume at STP

1 mole of CO_2 has 22.4 L volume at STP

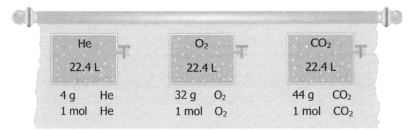

Example   2

What is the volume of 0.1 mol of hydrogen (H_2) gas at STP?

Solution

If 1 mol of H_2 22.4 L at STP then

0.1 mol of H_2 *x* L

$$x = \frac{0.1 \cdot 22.4}{1} = 2.24 \ L$$

Exercise 2:

What are the masses of 5.6 L of SO_3 and 5.6 L of CO gases at STP?

5. THE MOLE CONCEPT CALCULATIONS

The number of moles, which is represented by **n**, is used for various calculations in chemistry. The number of moles is related to the number of particles, mass or volume of substances.

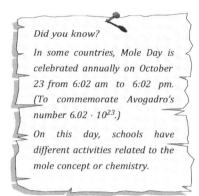

*In chemical calculations, use the **Periodic Table** in Appendix D to find the molar mass of any substance.*

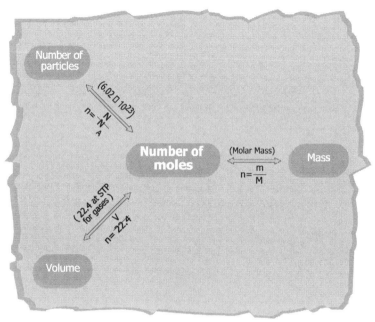

5.1. MOLE - NUMBER OF PARTICLES RELATIONSHIP

If the number of particles (atoms, molecules, ions....) for a substance is known, the mole number of that substance can easily be calculated, using the following:

$$\text{Number of moles} = \frac{\text{Number of particles}}{6.02 \cdot 10^{23}}$$

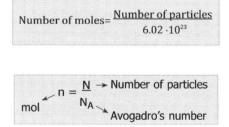

Example
3

What is the number of moles of $3.01 \cdot 10^{22}$ atoms of helium (He)?

Solution

By using the formula, $n_{He} = \dfrac{N}{N_A}$

$$n_{He} = \dfrac{3.01 \cdot 10^{22} \text{ helium atom}}{6.02 \cdot 10^{23}} = 0.05 \text{ mol}$$

Example
4

What is the total number of atoms found in two moles of a H_2SO_4 molecule?

Solution

By using the formula, $n_{(H_2SO_4)} = \dfrac{N}{N_A}$

$2 = \dfrac{N}{6.02 \cdot 10^{23}} \Rightarrow N = 1.204 \cdot 10^{24} \text{ H}_2\text{SO}_4 \text{ molecules}$

A H_2SO_4 molecule is composed of 7 atoms (2 **H**ydrogen, 1 **S**ulfur and 4 **O**xygen)

If a H_2SO_4 molecule contains   7 atoms

$1.204 \cdot 10^{24}$ H_2SO_4 molecules contain x

$$x = \dfrac{1.204 \cdot 10^{24} \cdot 7}{1} = 8.428 \cdot 10^{24} \text{ atoms}$$

Exercise 3:

If 10 moles of oxygen (O_2) molecules are given:

a) Calculate the number of oxygen molecules present.

b) Calculate the number of oxygen atoms present.

c) Calculate the volume of the oxygen molecules at STP.

Oxygen gas

5.2. MOLE - MASS RELATIONSHIP

The number of moles in a substance can be calculated with the help of its molar mass and mass. The following formula shows the relationship between the number of moles and mass.

Molar mass, MM or M, can be used for the atom, molecule or ion, etc...

$$\text{Number of moles} = \frac{\text{Mass}}{\text{Molar Mass}}$$

$$n = \frac{m}{M}$$ here $$\text{mol} \leftarrow n = \frac{m}{M} \rightarrow g \searrow g/mol$$

Example 5

What is the mole number of 11 g of carbon dioxide (CO_2) gas?

Solution

First, the molar mass of CO_2 is calculated. $M_{(CO_2)} = 1 \cdot 12 + 2 \cdot 16 = 44$ g/mol.

Then, the number of moles can be calculated by using the $n = \frac{m}{M}$ formula.

$$n = \frac{11g}{44g/mol} = 0.25 \text{ mol}$$

Example 6

What is the mass of 2.5 moles of sodium hydroxide (NaOH)?

Solution

$M_{(NaOH)} = (1 \cdot 23) + (1 \cdot 16) + (1 \cdot 1) = 40$ g/mol

Then, the formula mass of NaOH can be calculated by using $n = \frac{m}{M}$

$$2.5 \text{ mol} = \frac{m}{40 \text{ g/mol}} \Rightarrow m = 100 \text{ g}$$

Exercise 4:

How many moles of sulfur (S) and oxygen (O) exist in 32 g SO_2?

5.3. MOLE - VOLUME RELATIONSHIP

For gases at STP (0 °C and 1 atm), 1 mole of any gas occupies 22.4 L. Therefore, the number of moles of a gas can be calculated, if the volume of a gas is known at STP.

$$\text{Number of moles} = \frac{\text{Volume}}{22.4}$$

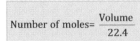

$$n = \frac{V}{22.4}$$

here

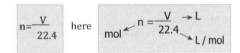

Example 7

What is the number of moles of 112 L chlorine (Cl_2) gas at STP?

Solution

By using the formula of $n = \dfrac{V}{22.4}$, the number of moles of Cl_2 can be calculated.

$$n = \frac{112 L}{22.4 L / mol}$$

$$n = 5 \text{ mol}$$

112 L Cl$_2$ n = ?

Example 8

How many liters does two moles of oxygen gas (O_2) occupy at STP?

Solution

$$n = \frac{V}{22.4} \Rightarrow 2 = \frac{V}{22.4} \Rightarrow V = 44.8 \text{ L}$$

2 mol O$_2$ V = ?

Exercise 5:

Answer the following questions for $3.01 \cdot 10^{23}$ carbon dioxide molecules (CO_2) at STP:

a) What is the volume of CO_2?

b) What is the total number of atoms in CO_2?

6. SEVERAL TYPES OF PROBLEMS

6.1. FINDING DENSITY OF A GAS AT STP

The density of gas at STP can be calculated if its molar mass is known.

Example 9

In general, the unit of density of a gas is given in g/L instead of g/mL or g/cm³.

Find the density of 2 moles of carbon dioxide (CO_2) at STP.

Solution

1 mole of CO_2 occupies	22.4 L at STP
2 mole of CO_2 occupies	x

$x = 44.8$ L

$M_{(CO_2)} = 12 + (16 \cdot 2) = 44$ g/mol.

Then the density of CO_2 is,

$$\rho = \frac{m}{V} \Rightarrow \frac{44\,g}{22.4\,L} = 1.96\,g/L$$

Exercise 6:

What is the density of oxygen (O_2) gas at STP?

6.2. MASS PERCENTAGE OF ELEMENTS IN A COMPOUND

The percentage of mass in an element in a compound can easily be found when the molecular formula and molar mass (or atomic mass) of elements are known.

Example 10

*Sodium hydroxide (NaOH) is known to be **caustic**.*

Find the mass percentages of each element in sodium hydroxide (NaOH).

Solution

$M_{(NaOH)} = 23 + 16 + 1 = 40$ g/mol

% **Na** in NaOH = ratio = $\dfrac{molar\ mass}{molar\ mass}$, % **O** in NaOH = $\dfrac{16}{40} \cdot 100 = 40\%$ and

% **H** in NaOH = $\dfrac{1}{40} \cdot 100 = 2.5\%$

Sucrose $(C_{12}H_{22}O_{11})$ *is better known as* **table sugar**.

Exercise 7:

What is the mass percentage of carbon (C) in sugar ($C_{12}H_{22}O_{11}$)?

6.3. CALCULATION OF EMPIRICAL FORMULA BY MASS PERCENTAGES

The mass of elements in a compound is crucial when solving problems with the empirical formula. First, the number of moles in atoms are found by dividing their masses by their molar masses. Next, the number of moles are either divided by the smallest mole number or, if necessary to calculate mole ratio of elements, multiplied with certain multipliers to get whole numbers for each element. These numbers for different elements provide the empirical formula.

Empirical Formula = Simplest Formula

Whole number = Integers (positive)
n = 0, 1, 2, 3 .....

Example 11

What is the empirical formula of 26.6 g nitrogen oxide (NO) that contains 9.8 grams of nitrogen?

$N_x O_y$

Solution

Total mass = $m_{(N)} + m_{(O)}$ = 26.6

$$26.6 = 9.8 + m_{(O)}$$

$$m_{(O)} = 16.8 \text{ g}$$

Step 1: Find the number of moles of atoms.

Number of moles of O : $n = \dfrac{16.8}{16} = 1.05$ mol,

Number of moles of N : $n = \dfrac{9.8}{14} = 0.7$ mol.

Step 2: Divide the number of moles by the smallest number, which is 0.7, to get the mole ratio between elements.

For N : $\dfrac{0.7}{0.7} = 1$,   for O : $\dfrac{1.05}{0.7} = 1.5$

Step 3: The number of atoms must be expressed as positive integers. Thus, the numbers should be multiplied by 2.

For N : $2 \cdot 1 = 2$, for O : $2 \cdot 1.5 = 3$

The empirical formula of the compound is N_2O_3.

Exercise 8:

A compound of C, H, and O is composed of 40 % of carbon, 6.66 % of hydrogen and 53.34 % of oxygen by mass. What is the empirical formula for this compound?

$C_x H_y O_z$

6.4. DETERMINING MOLECULAR FORMULA

The molecular formula of a compound can be calculated after its empirical formula has been found. First, the relation between the molar masses of molecular formula and the empirical formula is found by using the relation below.

$$ratio = \frac{\text{molar mass of molecular formula}}{\text{molar mass of empirical formula}}$$

The number of atoms in an empirical formula is multiplied by the obtained value in ratio calculations.

Example 12

Lactic acid is a compound (containing C, H and O), with a molar mass of 90 g/mol. 25.2 g of lactic acid contains 10.08 g of carbon and 13.44 g of oxygen. What is the molecular formula of lactic acid?

Solution

First, the mass of hydrogen in the compound is calculated.

25.2 – (10.08 + 13.44) = 1.68 g.

Then, we find the number of moles of C, H and O atoms.

Number of moles of C atoms, $n = \dfrac{10.08}{12} = 0.84$ mol.

Number of moles of H atoms, $n = \dfrac{1.68}{1} = 1.68$ mol.

Number of moles of O atoms, $n = \dfrac{13.44}{16} = 0.84$ mol.

Then, we divide these numbers by 0.84, $C_{\frac{0.84}{0.84}=1} \quad H_{\frac{1.68}{0.84}=2} \quad O_{\frac{0.84}{0.84}=1} \Rightarrow C_1H_2O_1$

The empirical formula for the compound is $C_1H_2O_1$ or $C_nH_{2n}O_n$.

The molar mass of $CH_2O = 12 + (2 \cdot 1) + 16 = 30$ g/mol.

$$ratio = \frac{\text{molar mass of lactic acid}}{\text{molar mass of empirical formula}} = \frac{90}{30} = 3 \text{ then}$$

The molecular formula of lactic acid is

$C_{1.3} \ H_{2.3} \ O_{1.3} \Rightarrow C_3H_6O_3$

Exercise 9:

What is the molecular formula, and the name, of 2 g of a compound that is composed of 0.5306 g K (potassium), 0.7076 g Cr (chromium) and 0.7618 g O (oxygen)?

A sample of potassium dichromate

Lactic acid is the acid found in milk.

6.5. MIXTURE PROBLEMS

In these types of problems, the components of a mixture can be calculated by using basic mathematical operations, as shown in the following example.

Example ——————————————————— 13

The mixture of oxygen and helium is 9.8 gram, and has a volume of 31.36 liters, at STP. What is the number of moles of each gas in the given mixture?

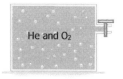

V = 31.36 L

Solution

First, find the number of moles in the mixture.

$$n_{(mixture)} = \frac{31.36\,L}{22.4\,L/mol} = 1.4 \text{ mol mixture}$$

If the number of oxygen in the mixture is accepted as x mole, then the number of moles of helium in the mixture becomes $1.4 - x$.

Remember the formula,

$$n = \frac{m}{M} \Rightarrow m = n \cdot M$$

Since the molar mass of O_2 is 32, and the molar mass of the He is 4; the masses of oxygen and helium gases are:

$m_{(O_2)} = 32 \cdot x$

$m_{(He)} = 4 \cdot (1.4 - x)$

and the value of x will be,

$$m_{(O_2)} + m_{(He)} = m_{(total)}$$
$$32 \cdot x + 4 \cdot (1.4 - x) = 9.8$$
$$x = 0.15 \text{ mol}$$

So, the number of moles of $O_{2(g)}$ is 0.15.

The number of moles of helium can be determined as such,

$n_{(He)} = 1.4 - x = 1.4 - 0.15 \Rightarrow n_{(He)} = 1.25 \text{ mol}$

Exercise 10:

What is the total mass of the mixture of C_3H_4, C_2H_2 and CH_4 gases, if each gas contains 2 moles of hydrogen atoms?

1. Why do we need Avogadro's number in chemistry?

2. Compare $6.02 \cdot 10^{23}$ (Avogadro's Number) and the population of the Earth, which is about 6.02 billions. How many planets (like Earth) would we need to populate to $6.02 \cdot 10^{23}$ people?

3. How can an atomic mass unit be converted into grams?

4. Find the molar masses for the following compounds (Look at the periodic table for the atomic mass of elements in **Appendix D**).

   a. CaO   **f.** $Na_2Cr_2O_7$

   b. $AlCl_3$ **g.** H_2SO_4

   c. NO_2 **h.** $(NH_4)_2SO_4$

   d. $Ca(OH)_2$

   e. NH_4NO_3

5. Find the masses for the following substances in grams.

   a. 1.5 mol of water

   b. 20 mol of C_2H_6

   c. 0.75 mol of H_2SO_4

6. Calculate how many atoms there are in each substance below. ($N_A = 6.02 \cdot 10^{23}$)

   a. 10 g of calcium

   b. 44.8 L of O_2 at STP

   c. 50 g of $CaCO_3$

7. Answer the following questions about 16 grams of SO_2 gas at STP.

   a. How many moles does it have?

   b. How much volume in liters does it occupy?

   c. How many molecules of SO_3 does it contain?

   d. How many atoms of S are present?

   e. How many grams of S does it contain?

8. Answer the following questions for 0.05 mol of H_2S?

   a. How many grams does it weigh?

   b. How many grams of hydrogen does it contain?

   c. How many grams of sulfur does it contain?

   d. How many molecules does it contain?

9. Find the percentage of nitrogen in the following compounds by mass?

   a. NH_3 **d.** NaCN

   b. HNO_3 **e.** N_2O_3

   c. NH_4NO_3 **f.** N_2O_5

10. Calculate the percentage of composition for each element in the following compounds by mass.

   a. SO_2 **b.** $CaCO_3$ **c.** MgO   **d.** H_2SO_4

11. At STP, 5.6 L of XO_2 gas weighs 11 g. Calculate the molar mass of X.

12. Find the empirical formula for the following compounds containing:

   a. Compound 60 % magnesium and 40 % oxygen

   b. Compound 30 % oxygen and 70 % iron

   c. Compound 50 % copper and 50 % sulfur

13. 0.5 mol of a compound containing Ca, C and O atoms is 50 g. If this compound contains 40 % of calcium, 12 % of carbon and 48 % of oxygen, what is the molecular formula of the compound?

14. An 11.2 L mixture of He and O_2 gases weighs 6.2 g at STP. What is the percentage of mol of O_2 gas in the mixture?

MULTIPLE CHOICE QUESTIONS

1. Which of the following is not true for $6.02 \cdot 10^{23}$?

   A) It is called Avogadro's number.

   B) It is abbreviated as N_A.

   C) 1 mol of any substance contains that amount of particles.

   D) It is a number of atoms found in the Earth.

   E) It is a well known number for chemists.

2. Which of the following elements or compounds has 1 mol of atoms?

   A) $1.204 \cdot 10^{22}$ O_2 molecules

   B) $1.204 \cdot 10^{22}$ CO_2 molecules

   C) $1.204 \cdot 10^{23}$ HNO_3 molecules

   D) $1.204 \cdot 10^{24}$ H_2O molecules

   E) $1.204 \cdot 10^{25}$ H_2 molecules

3. How many moles of atoms are there in 0.5 mol of P_4O_{10}?

   A) 3   B) 4   C) 5   D) 6   E) 7

4. What is the volume of $1.204 \cdot 10^{22}$ of CO_2 molecules in mL at STP?

   A) 0.448   B) 4.48   C) 44.8   D) 448   E) 4488

5. Which of the following compounds has the highest mass value?

   A) 4 mol of H_2

   B) 0.2 mol of CO_2

   C) $6.02 \cdot 10^{24}$ of H_2O molecules

   D) 2.24 L of C_4H_{10} gas at STP

   E) 10 g of CO_2

6. What is the mass of a water (H_2O) molecule in grams? ($N_A \cong 6 \cdot 10^{23}$)

   A) $6 \cdot 10^{-22}$ B) $3 \cdot 10^{-23}$ C) $6 \cdot 10^{-23}$
   D) $3 \cdot 10^{-24}$ E) $6 \cdot 10^{-24}$

7. What is the formula of an oxide of sulfur, if it contains 40% of sulfur and 60% of oxygen by mass?

   A) SO B) SO_2 C) S_2O_5 D) S_2O_3 E) SO_3

8. What is the ratio of density of O_2 gas to density of air? (1 mol of air is approximately 29 g/mol at STP)

   A) 16/29   B) 32/29   C) 29/32   D) 32/1   E) 1/29

9. What is the value of n, if 0.5 mol of C_nH_{2n} is 21 grams?

   A) 1   B) 2   C) 3   D) 4   E) 5

10. At 16 grams, which of the following compounds has the largest volume at STP?

    A) O_2 B) SO_2 C) CO_2 D) CH_4 E) NO_2

CRISSCROSS PUZZLE

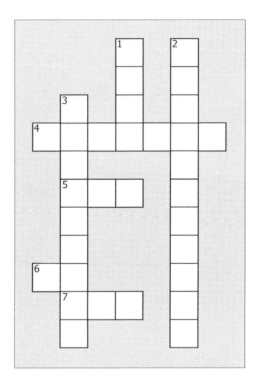

ACROSS

  4 An annual celebration day for chemists

  5 Atomic mass unit

  6 Avogadro's number

  7 Standard temperature and pressure

DOWN

  1 A chemical unit

  2 Volume of one mole of gas at standard conditions

  3 Mass of one mole of substance

APPENDICES
GLOSSARY
ANSWERS
INDEX

INTRODUCTION
TO CHEMISTRY

APPENDICES

Appendix A

Some quantities and units

Physical quantity	Units
Area	cm^2, m^2
Volume	cm^3, dm^3, m^3, litre (L), millimetre (mL)
Density	kg/m^3, g/cm^3
Force	newton (N)
Pressure	pascal (Pa or N/m^2), N/cm^2
Speed	m/s, km/h
Acceleration	m/s^2
Energy	joule (J), kilojoule (kJ), megajoule (MJ)
Power	watt (W), kilowatt (kW), megawatt (MW)
Frequency	hertz (Hz), kilohertz (kHz)
Electrical charge	coulomb (C)
Potential difference	volt (V)
Resistance	ohm (Ω)
Gravitational field strength	N/kg
Radioactivity	becquerel (Bq)
Sound intensity	decibel (dB)

Appendix B

Conversion factors

Factor	Name	Symbol	Factor	Name	Symbol
10^{24}	yotta	Y	10^{-1}	deci	d
10^{21}	zetta	Z	10^{-2}	centi	c
10^{18}	exa	E	10^{-3}	milli	m
10^{15}	peta	P	10^{-6}	micro	μ
10^{12}	tera	T	10^{-9}	nano	n
10^{9}	giga	G	10^{-12}	pico	p
10^{6}	mega	M	10^{-15}	femto	f
10^{3}	kilo	k	10^{-18}	atto	a
10^{2}	hecto	h	10^{-21}	zepto	z
10^{1}	deca	da	10^{-24}	yacto	y

Appendix C

Three well – known temperature scales and their relationship

	Celsius	Kelvin	Fahrenheit
Absolute zero	-273	0	-460
Freezing point of water (at standard pressure)	0	273	32
Average body temperature	$36.5 - 37.2$	310	98
Boiling point of water	100	373	212

$$°C = [°F - 32] \cdot \frac{5}{9}$$

$$°F = \frac{9}{5} \cdot °C + 32$$

$$K = °C + 273$$

Appendix D

The Pictorial Periodic Table of Elements

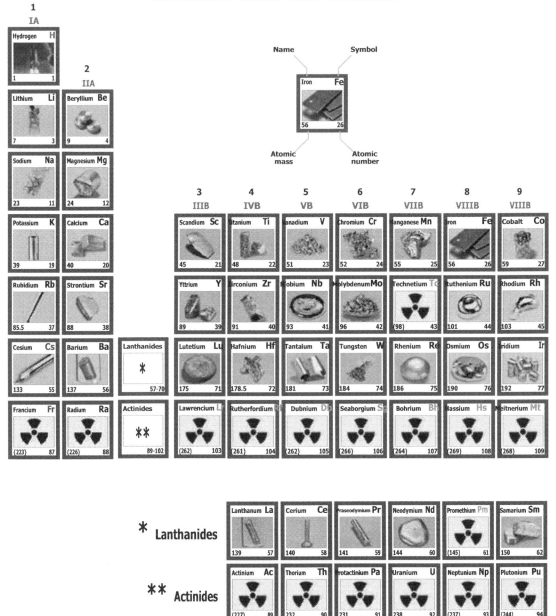

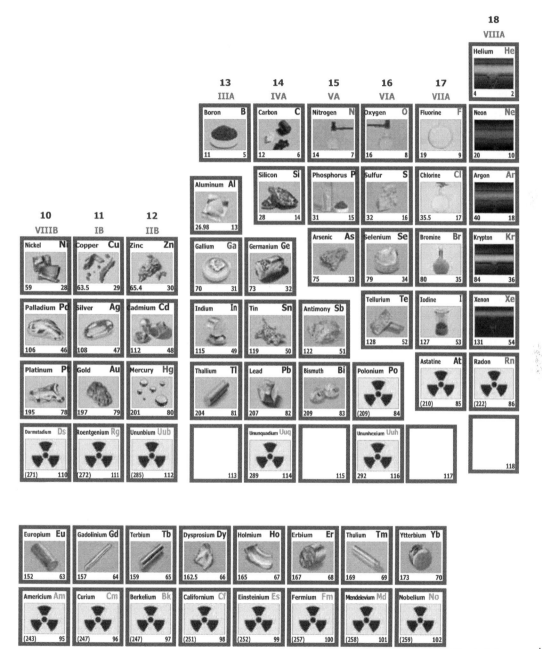

GLOSSARY

Atomic mass : The mass of an atom in amu (atomic mass unit).

Atomic mass unit : 1/12 (one-twelfth) of the mass of a carbon - 12 atom.

Atomic number : Number of protons in an atom.

Avogadro's number : $6.02 \cdot 10^{23}$.

Atom : The tiny particle that makes up all substances.

Anion : Negatively charged ion.

Average atomic mass : Average of atomic masses of isotopes by their fractional abundance.

Biology : A branch of natural science. Biology is the study of living organisms.

Boiling point : The temperature at which the vapor pressure of a liquid is equal to the external (air) pressure.

Cation : Positively charged ion.

Chemical property : A property that changes the chemical nature of matter.

Chemical change : A change that occurs when chemical properties of a substance change.

Chemist : Scientist who studies chemistry

Compound : A pure substance made up of two or more different elements.

Chemistry : A branch of natural science that studies matter and its changes.

Condensation : Process in which liquids become gases.

Covalent bond : The bond in which two electrons are shared by two atoms (bonds between nonmetal atoms).

Density : The mass of a unit volume of a substance.

Distillation : A process for separation of mixtures through their different boiling points.

Experiment : An organized procedure to perform the processes under controlled conditions.

Extensive property : A property that depends on the amount of matter.

Electron : Negatively charged subatomic particle.

Element : A pure substance that cannot be broken down into simpler elements.

Evaporation : Process in which gas becomes liquid.

Freezing : Process in which liquid becomes solid.

Heterogeneous mixtures : A type of mixture of two or more substances that are not uniformly distributed throughout the mixture.

Homogeneous mixture : A type of mixture of two or more substances that are uniformly distributed throughout the mixture.

Ion : Charged atom.

Intensive property : A property that does not depend on the amount of matter.

Ionic bond : The bond formed between a metal and a nonmetal as a result of an electron transfer.

Isotopes : Atoms that share the same number of protons but different numbers of neutrons.

Law of definite proportion : A chemical law that shows the relation between elements in a compound.

Law of multiple proportion : A chemical law that shows the relation between elements that have more than one compound.

Laboratory : Special place for conducting experiments.

Mass : Amount of substance.

Matter : Anything that has mass and volume.

Measurement : Comparison of a quantity with a standard unit.

Mixture : Combination of two or more substances in any amount.

Mole : A unit of chemistry, that shows the $6.02 \cdot 10^{23}$ proportion of any particle.

Molecular mass : Mass of a molecule.

Molecule : Basic unit for molecular (covalent) compounds.

Molecular formula : Formula of a compound (molecular).

Molar mass : Mass of a substance.

Melting point : The temperature at which solids start to liquify.

Neutron : A non-charged subatomic particle found in the nucleus.

Neutral atom : An atom which has the same number of protons and electrons.

Nucleus (pl. nuclei) : The minute, very dense positively charged center of an atom.

Orbitals : Places of electrons (different energy levels).

Oxidation state : Valency of an atom.

Periodic Table : Table of elements.

Plasma : The fourth state of matter.

Physical property : A property of a substance that can be observed and measured without changing it into another substance.

Proton : A positively charged particle found in the nucleus of an atom.

Relative molecular mass : Some of the average atomic masses found in a compound.

STP : Standard temperature and pressure (1 atm and °C).

Solute : The substance in a solution that is dissolved by a solvent.

Solvent : The substance that dissolves a solute in a solution.

Solution : *See* homogeneous mixture.

Temperature : Degree of hotness or coldness.

Technology : Use of accumulated knowledge (science) to produce new substances, or processes.

Valence electrons : Electrons in the outermost shell of an atom.

Volume : The space occupied by substances.

ANSWERS

CHEMISTRY A UNIQUE SCIENCE

15. a. =

b. >

c. =

d. <

16. 5 mL = 5 cm^3

17. 1. b

2. a

3. d

4. e

5. c

18. a. Test tube

b. Graduated cylinder

c. Erlenmeyer flask

21. a. 54.75 ton

b. 54750 L

c. No

MATTER

5. a. Mixture

b. Mixture

c. Pure Substance

d. Mixture

e. Mixture

f. Pure Substance

g. Mixture

h. Mixture

i. Mixture

j. Pure Substance

k. Mixture

l. Pure Substance

8. a. Helium   b. Uranium

c. Silver   d. Nitrogen

13. 20 cm^3

18. a. 2   b. 2   c. 2

ATOM

4. a. Al: 2) 8) 3)

b. K: 2) 8) 8) 1)

9.

Element	p	n	charge	e$^-$	A
Mg	**12**	12	**+2**	10	24
Cl	**17**	18	−1	**18**	35
S	16	**16**	**0**	16	32
Na	11	12	+1	**10**	**23**

10. X and Z

12. a. $_{8}^{16}\text{O}_{8}^{2-}{}_{10}$ b. $_{30}^{56}\text{Fe}_{26}^{2+}{}_{24}$

c. $_{20}^{40}\text{Ca}_{20}^{2+}{}_{18}$ d. $_{108}^{197}\text{Au}_{79}^{3+}{}_{76}$

14. 20

COMPOUNDS

5. a. CaO
    b. Al_2S_3
    c. MgF_2
    d. $FeCl_2$
    e. SnO_2

6. a. Silver chloride
    b. Sodium bromide
    c. Copper (I) oxide
    d. Zinc sulfide
    e. Hydrogen chloride

7. a. $KClO_3$
    b. $Al_2(SO_4)_3$
    c. HCN
    d. $AgNO_3$
    e. $Zn(OH)_2$

8. a. Magnesium hydroxide
    b. Sodium chromate
    c. Calcium nitrate
    d. Iron (II) sulfate
    e. Ammonium phosphate

9. a. N_2O
    b. OF_2
    c. XeF_4
    d. PCl_5

10. a. Nitrogen dioxide
    b. Iodine pentafluoride
    c. Chlorine trifluoride
    d. Sulfur tetrafluoride

MOLE CONCEPT

2. 10^{14} planets!

4. a. 56
    b. 133.5
    c. 46
    d. 74
    e. 80
    f. 262
    g. 98
    h. 132

5. a. 27
    b. 60
    c. 73.5

6. a. $1.204 \cdot 10^{23}$
    b. $2.408 \cdot 10^{24}$
    c. $1.505 \cdot 10^{24}$

7. a. 0.25 mol
    b. 5.6 L
    c. $1.505 \cdot 10^{23}$ molecules
    d. $1.505 \cdot 10^{23}$ S atoms
    e. 8 g

8. a. 1.7 g
    b. 0.1 g
    c. 1.6 g
    d. $3.01 \cdot 10^{22}$

9. a. 82.35%
    b. 22.22%
    c. 35.00%
    d. 28.57%
    e. 36.84%
    f. 25.93%

10. a. S: 50%, O:50%
    b. Ca: 40%, C: 12%, O: 48%
    c. Mg: 42.86%, O: 57.14%
    d. H: 2.04%, S:32.65, O: 65.31

11. 12g

12. a. MgO
    b. Fe_2O_3
    c. CuO_2

13. $CaCO_3$

14. 30%

MULTIPLE CHOICE

CHEMISTRY, A UNIQUE SCIENCE

1. C	4. C	7. D	10. E
2. C	5. C	8. E	11. A
3. D	6. A	9. E	12. B

MATTER

1. D	4. E	7. D	10. B
2. A	5. C	8. B	
3. C	6. A	9. C	

ATOM

1. C	4. D	7. C	10. D
2. D	5. E	8. B	
3. A	6. A	9. D	

COMPOUNDS

1. C	4. A	7. D	10. B
2. B	5. C	8. D	
3. A	6. D	9. E	

MOLE CONCEPT

1. B	4. A	7. E	10. D
2. C	5. C	8. B	
3. E	6. B	9. C	

EXERCISES

1. a. 1   b. 3

2. b and c

3. 45

4. 8

5. 64 amu and 101 amu

COMPOUNDS

1. a. Na_2O b. CaS   c. AlF_3

2. +4

3. $AlCl_3$ and KI

4. a. Potassium permanganate
    b. Aluminum hydroxide

5. a. $Al_2(SO_4)_3$: Aluminum sulfate
    b. $Al_2(CO_3)_3$: Aluminum carbonate
    c. $AlPO_4$: Aluminum phosphate

6. a. Lead (IV) chloride
    b. Lead (II) chloride

7. a. Dinitrogen tetroxide
    b. Carbon dioxide
    c. Sulfur hexafluoride
    d. Dinitrogen pentoxide
    e. Diphosphorus trioxide
    f. Carbon disulfide

8. $S/O = 2/3$

MOLE CONCEPT

1. 342 g

2. 20 g and 7 g

3. a. $6.02 \cdot 10^{24}$
    b. $1.204 \cdot 10^{25}$
    c. 224 L

4. 0.5 mol S and 1 mol O

5. a. 11.2 L
    b. $9.03 \cdot 10^{23}$ atoms

6. 1.43 g/L

7. 42.1%

8. CH_2O

9. $K_2Cr_2O_7$ and potassium dichromate

10. 54 g

PUZZLE

CHEMISTRY, A UNIQUE SCIENCE

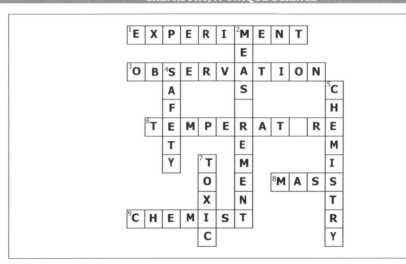

MATTER

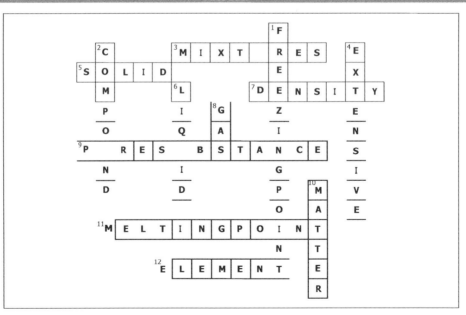

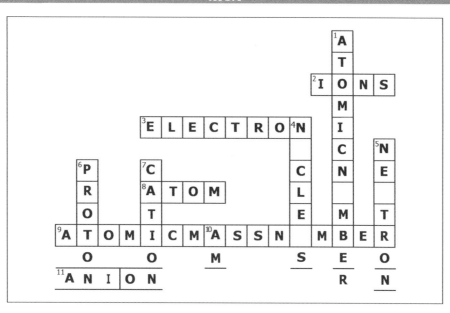

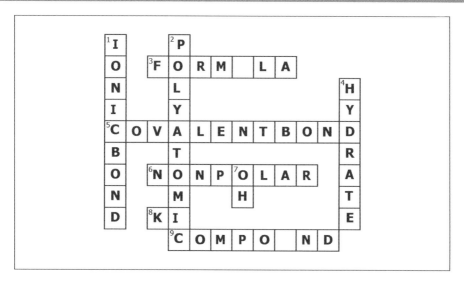

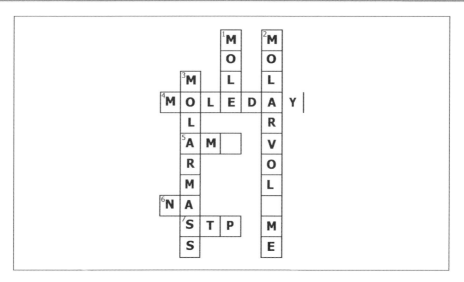

INDEX

REFERENCES

Çelik, N. **Chemistry 1 : As a Popular Science.** Sürat Publishing, Istanbul: 2002

Çelik, N., Erdem, A.R., Gürler, V., Karabürk, H., Nazlý, P. **Chemistry 2 : As a Popular Science.** Sürat Publications, Istanbul : 2000

Nazlý, A., Bayar, M., Duran, C., Çelik, N., Karabürk, H., Patlý, U.H. **Lise Kimya 1. Chemistry for Lycée 1.** Zambak Publication, Istanbul : 2004

Nazlý, A, **Chemistry : Laboratory Experiments.** Zambak Publication, Istanbul : 2003

Frank, D.V., Little, J.G., Miller, S. **Chemical Interactions.** Prentice Hall Science Explorer, Pearson Education Inc., New Jersey : 2005

Wilson, D., Bauer, M. **Dynamic Science.** McGraw-Hill., Sydney : 1995

Haire, M.; Kennedy, E., Lofts, G., Evergreen, M.J. **Core Science 2.** John Wiley & Sons Australia Ltd., Sydney : 1999

Oxtoby, D.W. Nachtrieb, N.H. **Principles of Modern Chemistry, 3rd Edition.** Saunders College Publishing, USA : 1996

Sevenair, J.P., Burkett, A.R. **Introductory Chemistry : Investigating the Molecular Nature of Matter.** WCB Publishers, USA : 1997

Prescott, C.N. **Chemistry : A Course for "O" Level.** Times Media Private Limited, Singapore : 2000

Gallagher. RM., Ingram, P. **Modular Science, Chemistry.** Oxford University Press, UK : 2001

Ryan, L. **Chemistry for You.** Stanley Thornes (Publishers) Ltd. UK : 1996

Lidin, R.A., Molochko B.A., Andreyeva L.L.; **Khimicheski Svoytsva Neorganicheskih Veshestv, 3rd edition.** Khimia, Moscow: 2000

Masterton, W.L. and Hurley, C.N. **Chemistry : Principles and Reactions, 3rd edition.** Saunders College Publishing, USA : 1996

Tsetkovik, S. **Khemija.** Prosvetno Dela Ad. Skopje : 2002

Printed in Great Britain
by Amazon